GCSE

Chemistry

Complete Revision and Practice

Contents

Contents

The Periodic Table

Periods

Group 0

1

| 1 H Hydrogen 1 | | | | | | | | | | | | | | | | 4 He Helium 2 |

Group 1 — Group 2

Group 3 — Group 4 — Group 5 — Group 6 — Group 7

Atomic number → 5

2

| 7 Li Lithium 3 | 9 Be Beryllium 4 | | | | | | | | | | 11 B Boron 5 | 12 C Carbon 6 | 14 N Nitrogen 7 | 16 O Oxygen 8 | 19 F Fluorine 9 | 20 Ne Neon 10 |

3

| 23 Na Sodium 11 | 24 Mg Magnesium 12 | | | | | | | | | | 27 Al Aluminium 13 | 28 Si Silicon 14 | 31 P Phosphorus 15 | 32 S Sulfur 16 | 35.5 Cl Chlorine 17 | 40 Ar Argon 18 |

4

| 39 K Potassium 19 | 40 Ca Calcium 20 | 45 Sc Scandium 21 | 48 Ti Titanium 22 | 51 V Vanadium 23 | 52 Cr Chromium 24 | 55 Mn Manganese 25 | 56 Fe Iron 26 | 59 Co Cobalt 27 | 59 Ni Nickel 28 | 63.5 Cu Copper 29 | 65 Zn Zinc 30 | 70 Ga Gallium 31 | 73 Ge Germanium 32 | 75 As Arsenic 33 | 79 Se Selenium 34 | 80 Br Bromine 35 | 84 Kr Krypton 36 |

5

| 85 Rb Rubidium 37 | 88 Sr Strontium 38 | 89 Y Yttrium 39 | 91 Zr Zirconium 40 | 93 Nb Niobium 41 | 96 Mo Molybdenum 42 | 99 Tc Technetium 43 | 101 Ru Ruthenium 44 | 103 Rh Rhodium 45 | 106 Pd Palladium 46 | 108 Ag Silver 47 | 112 Cd Cadmium 48 | 115 In Indium 49 | 119 Sn Tin 50 | 122 Sb Antimony 51 | 128 Te Tellurium 52 | 127 I Iodine 53 | 131 Xe Xenon 54 |

6

| 133 Cs Caesium 55 | 137 Ba Barium 56 | 57-71 Lanthanides | 178 Hf Hafnium 72 | 181 Ta Tantalum 73 | 184 W Tungsten 74 | 186 Re Rhenium 75 | 190 Os Osmium 76 | 192 Ir Iridium 77 | 195 Pt Platinum 78 | 197 Au Gold 79 | 201 Hg Mercury 80 | 204 Tl Thallium 81 | 207 Pb Lead 82 | 209 Bi Bismuth 83 | 210 Po Polonium 84 | 210 At Astatine 85 | 222 Rn Radon 86 |

7

| 223 Fr Francium 87 | 226 Ra Radium 88 | 89-103 Actinides | | | | | | | | | | | | | | | |

Published by CGP

From original material by Richard Parsons.

Editors:
Katherine Craig, Emma Elder, Mary Falkner, Ceara Hayden, Camilla Simson, Karen Wells.

Contributors:
Michael Aicken, Mike Bossart, Mike Dagless, Max Fishel, Gemma Hallam, Andy Rankin, Mike Thompson, Chris Workman

ISBN: 978 1 84146 658 3

With thanks to Rosie McCurrie and Helen Ronan for the proofreading.
With thanks to Anna Lupton for the copyright research.

Graph to show trend in atmospheric CO_2 concentration and global temperature on page 93 based on data by EPICA Community Members 2004 and Siegenthaler et al 2005.

Page 144 contains public sector information published by the Health and Safety Executive and licensed under the Open Government Licence v1.0.
http://www.nationalarchives.gov.uk/doc/open-government-licence/

Groovy website: www.cgpbooks.co.uk

Printed by Elanders Ltd, Newcastle upon Tyne.
Jolly bits of clipart from CorelDRAW®

The Scientific Process

Before you get started with the really fun stuff, it's a good idea to understand exactly how the world of science works. Investigate these next few pages and you'll be laughing all day long on results day.

Scientists Come Up with **Hypotheses** — Then **Test** Them

1) Scientists try to explain things. Everything.

2) They start by observing or thinking about something they don't understand — it could be anything, e.g. planets in the sky, a person suffering from an illness, what matter is made of... anything.

3) Then, using what they already know (plus a bit of insight), they come up with a hypothesis — a possible explanation for what they've observed.

4) The next step is to test whether the hypothesis might be right or not — this involves gathering evidence (i.e. data from investigations).

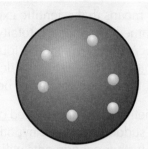

About 100 years ago, scientists hypothesised that atoms looked like this.

5) To gather evidence the scientist uses the hypothesis to make a prediction — a statement based on the hypothesis that can be tested by carrying out experiments.

6) If the results from the experiments match the prediction, then the scientist can be more confident that the hypothesis is correct. This doesn't mean the hypothesis is true though — other predictions based on the hypothesis might turn out to be wrong.

Scientists **Work Together** to Test Hypotheses

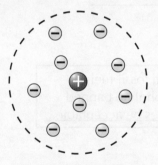

Then they thought they looked like this.

1) Different scientists can look at the same evidence and interpret it in different ways. That's why scientists usually work in teams — they can share their different ideas on how to interpret the data they find.

2) Once a team has come up with (and tested) a hypothesis they all agree with, they'll present their work to the scientific community through journals and scientific conferences so it can be judged — this is called the peer review process.

3) Other scientists then check the team's results (by trying to replicate them) and carry out their own experiments to collect more evidence.

4) If all the experiments in the world back up the hypothesis, scientists start to have a lot of confidence in it. (A hypothesis that is accepted by pretty much every scientist is referred to as a theory.)

5) However, if another scientist does an experiment and the results don't fit with the hypothesis (and other scientists can replicate these results), then the hypothesis is in trouble. When this happens, scientists have to come up with a new hypothesis (maybe a modification of the old explanation, or maybe a completely new one).

A hypothesis is a possible explanation for an observation

If scientists think something is true, they need to produce evidence to convince others — it's all part of testing a hypothesis. One hypothesis might survive these tests, while others won't — it's how things progress. And along the way some hypotheses will be disproved — i.e. shown not to be true.

The Scientific Process

Scientific Ideas **Change** as **New Evidence** is Found

1) Scientific explanations are <u>provisional</u> because they only explain the evidence that's <u>currently available</u> — new evidence may come up that can't be explained.

2) This means that scientific explanations <u>never</u> become hard and fast, totally indisputable <u>fact</u>. As <u>new evidence</u> is found (or new ways of <u>interpreting</u> existing evidence are found), hypotheses can <u>change</u> or be <u>replaced</u>.

3) Sometimes, an <u>unexpected observation</u> or <u>result</u> will suddenly throw a hypothesis into doubt and further experiments will need to be carried out. This can lead to new developments that <u>increase</u> our <u>understanding</u> of science.

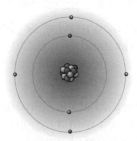

Then they thought they looked like this.

Scientific Developments are **Great**, but they can **Raise Issues**

Scientific <u>knowledge is increased</u> by doing experiments.
And this knowledge leads to <u>scientific developments</u>, e.g. new technologies or new advice.
These developments can create <u>issues</u> though. For example:

Economic issues:

> Society <u>can't</u> always <u>afford</u> to do things scientists recommend (e.g. investing heavily in alternative energy sources) without <u>cutting back elsewhere</u>.

Social issues:

> Decisions based on scientific evidence affect <u>people</u> — e.g. should fossil fuels be taxed more highly (to invest in alternative energy)? Should alcohol be banned (to prevent health problems)? <u>Would the effect on people's lifestyles be acceptable...</u>

Environmental issues:

> <u>Chemical fertilisers</u> may help us <u>produce more food</u> — but they also cause <u>environmental problems</u>.

Ethical issues:

> There are a lot of things that scientific developments have made possible, but <u>should we do them</u>? E.g. clone humans, develop better nuclear weapons.

It's not all test tubes and explosions

Life can be hard as a scientist. You think you've got it all figured out and then someone comes along with some <u>issues</u>. But it's for the best really — the world would be pretty messed up if no-one ever thought about the issues created by scientific developments (just watch Jurassic Park...).

Quality of Data

When it comes to evidence, <u>reliability</u> and <u>validity</u> are really important.

Evidence Needs to be Reliable (Repeatable and Reproducible)

<u>RELIABLE</u> means that the data can be <u>repeated</u>, and <u>reproduced</u> by others.

Evidence is only <u>reliable</u> if it can be <u>repeated</u> (during an experiment) AND <u>other scientists can reproduce it too</u> (in other experiments). If it's not reliable, you can't believe it.

<u>Example: Cold fusion</u>

- In 1989, two scientists claimed that they'd produced '<u>cold fusion</u>' (the energy source of the Sun — but without the big temperatures).

- It was huge news — if true, it would have meant cheap and abundant energy for the world... forever.

- However, other scientists just <u>couldn't reproduce the results</u> — so the results <u>weren't reliable</u>. And until they are, 'cold fusion' isn't going to be accepted as <u>fact</u>.

Evidence Also Needs to Be Valid

<u>VALID</u> means that the data is <u>reliable</u> AND <u>answers the original question</u>.

<u>Example: Do power lines cause cancer?</u>

- Some studies have found that children who live near <u>overhead power lines</u> are more likely to develop <u>cancer</u>. What they'd actually found was a <u>correlation</u> (relationship) between the variables "<u>presence of power lines</u>" and "<u>incidence of cancer</u>" — they found that as one changed, so did the other.

- But this evidence is <u>not enough</u> to say that the power lines <u>cause</u> cancer, as other explanations might be possible. For example, power lines are often near <u>busy roads</u>, so the areas tested could contain <u>different levels</u> of <u>pollution</u> from traffic.

- So these studies don't show a definite link and so don't <u>answer the original question</u>.

RRRR — Remember, Reliable means Repeatable and Reproducible

The scientific community <u>won't accept</u> someone's data if it can't be <u>repeated</u> by anyone else. It may sound like a really fantastic new theory, but if there's no other <u>support</u> for it, it just <u>isn't reliable</u>.

Quality of Data

The way evidence is <u>gathered</u> can have a big effect on how <u>trustworthy</u> it is...

The **Bigger** the **Sample Size** the **Better**

1) Data based on <u>small samples</u> isn't as good as data based on large samples.

> A sample should be <u>representative</u> of the <u>whole</u>
> <u>population</u> (i.e. it should share as many of the
> various characteristics in the population as
> possible) — a small sample can't do that as well.

2) The <u>bigger</u> the sample size the <u>better</u>, but scientists have to be <u>realistic</u> when choosing how big.

> For example, if you were studying how lifestyle affects
> people's weight it'd be great to study everyone in
> the UK (a huge sample), but it'd take <u>ages</u> and <u>cost a</u>
> <u>bomb</u>. Studying a thousand people is more <u>realistic</u>.

Don't Always **Believe** What You're Being **Told Straight Away**

1) People who want to make a point might <u>present data</u> in a <u>biased way</u>, e.g. by overemphasising a relationship in the data. (Sometimes <u>without knowing</u> they're doing it.)

2) And there are all sorts of reasons <u>why</u> people might <u>want</u> to do this — for example, <u>companies</u> might want to 'big up' their products. Or make impressive safety claims.

3) If an investigation is done by a team of <u>highly-regarded scientists</u> it's sometimes taken <u>more seriously</u> than evidence from <u>less well known scientists</u>.

4) But having experience, authority or a fancy qualification <u>doesn't</u> necessarily mean the evidence is <u>good</u> — the only way to tell is to look at the evidence scientifically (e.g. is it reliable, valid, etc.).

Things are not always what they seem

No matter <u>what</u> you're reading or <u>who</u> it's written by you've always got to be really careful about what you <u>believe</u>. Ask yourself whether the <u>sample</u> is a decent <u>size</u> and check whether the author has anything to gain from what's written. For example, an article on the magical fat-busting power of spinach written by the country's leading spinach grower may not be all it seems. <u>Don't be fooled</u>.

Limits of Science

Science can give us amazing things — cures for diseases, space travel, heated toilet seats...
But science has its limitations — there are questions that it just can't answer.

Some Questions Are *Unanswered*...

Some questions are unanswered — we don't know everything and we never will.
We'll find out more as new hypotheses are suggested and more experiments are done,
but there'll always be stuff we don't know.

> For example, we don't know what the exact impacts of global warming are going to be. At the moment scientists don't all agree on the answers because there isn't enough reliable and valid evidence.

...Others are *Unanswerable*

1) Then there's the other type... questions that all the experiments in the world won't help us answer — the "Should we be doing this at all?" type questions. There are always two sides...

2) Think about new drugs which can be taken to boost your 'brain power'.

3) Different people have different opinions. For example...

Some people think they're good... Or at least no worse than taking vitamins or eating oily fish. They could let you keep thinking for longer, or improve your memory. It's thought that new drugs could allow people to think in ways that are beyond the powers of normal brains — in effect, to become geniuses...

Other people say they're bad... taking them would give you an unfair advantage in exams, say. And perhaps people would be pressured into taking them so that they could work more effectively, and for longer hours.

4) The question of whether something is morally or ethically right or wrong can't be answered by more experiments — there is no "right" or "wrong" answer.

5) The best we can do is get a consensus from society — a judgement that most people are more or less happy to live by. Science can provide more information to help people make this judgement, and the judgement might change over time. But in the end it's up to people and their conscience.

To answer or not to answer, that is the question

It's official — no-one knows everything. Your teacher/mum/annoying older sister (delete as applicable) might think and act as if they know it all, but sadly they don't. So in reality you know one thing they don't — which clearly makes you more intelligent and generally far superior in every way. Possibly.

Planning Investigations

The next few pages show how <u>investigations</u> should be carried out — by both <u>scientists</u> and <u>you</u>.

In a *Fair Test* You Have to *Control the Variables*

1) In a lab experiment you usually <u>change one variable</u> and <u>measure</u> how it affects the <u>other variable</u>.

   EXAMPLE: you might change only the temperature of a chemical reaction
   and measure how this affects the rate of reaction.

2) To make it a fair test <u>everything else</u> that could affect the results should <u>stay the same</u>
   (otherwise you can't tell if the thing that's being changed is affecting the results or not
   — the data won't be reliable or valid).

   EXAMPLE continued: you need to keep the concentration of the reactants the
   same, otherwise you won't know if any change in the rate of reaction is caused
   by the change in temperature, or a difference in reactant concentration.

3) The variable that you <u>change</u> is called the <u>independent</u> variable.

4) The variable that's <u>measured</u> is called the <u>dependent</u> variable.

5) The variables that you <u>keep the same</u> are called <u>control</u> variables.

   EXAMPLE continued:
   Independent = temperature
   Dependent = rate of reaction
   Control =
   reactant concentration

6) Because you can't always control all the variables, you often
   need to use a <u>control experiment</u> — an experiment that's kept
   under the <u>same conditions</u> as the rest of the investigation, but doesn't have anything done
   to it. This is so that you can see what happens when you don't change anything at all.

Experiments Must be *Safe*

1) Part of planning an investigation is making sure that it's <u>safe</u>.

2) A <u>hazard</u> is something that can <u>potentially cause harm</u>.

3) There are lots of <u>hazards</u> you could be faced with during an investigation,
   e.g. <u>radiation</u>, <u>electricity</u>, <u>gas</u>, <u>chemicals</u> and <u>fire</u>.

4) You should always make sure that you <u>identify</u> all the hazards
   that you might encounter.

5) You should also come up with ways of <u>reducing the risks</u> from the hazards you've identified.

6) One way of doing this is to carry out a <u>risk assessment</u>:

   For an experiment
   involving a <u>Bunsen burner</u>,
   the risk assessment might
   be something like this:

   <u>Hazard:</u> Bunsen burner is a fire risk.
   <u>Precautions:</u>
   • Keep flammable chemicals away from the Bunsen.
   • Never leave the Bunsen unattended when lit.
   • Always turn on the yellow safety flame when not in use.

Hazard: revision boredom. Precaution: use CGP books

Labs are dangerous places — you need to know the <u>hazards</u> of what you're doing <u>before you start</u>.

Collecting Data

There are a few things that can be done to make sure that you get the <u>best results</u> you possibly can.

The **Equipment** Used has to be **Right for the Job**

1) The measuring equipment you use has to be <u>sensitive enough</u> to accurately measure the chemicals you're using.

E.g. if you need to measure out 11 ml of a liquid, you'll need to use a measuring cylinder that can measure to 1 ml, not 5 or 10 ml.

2) The <u>smallest change</u> a measuring instrument can <u>detect</u> is called its <u>RESOLUTION</u>. E.g. some mass balances have a resolution of 1 g and some have a resolution of 0.1 g.

3) Also, equipment needs to be <u>calibrated</u> so that your data is <u>more accurate</u>. E.g. mass balances need to be set to zero before you start weighing things.

Accurate data is data that's close to the true values — see the next page.

Trial Runs give you the Range and Interval of Variable Values

1) Before you carry out an experiment, it's a good idea to do a <u>trial run</u> first — a <u>quick version</u> of your experiment.

2) Trial runs help you work out whether your plan is <u>right or not</u> — you might decide to make some <u>changes</u> after trying out your method.

3) Trial runs are used to figure out the <u>range</u> of variable values used (the upper and lower limit).

<u>Reaction example from previous page continued</u>:
You might do trial runs between 10 and 60 °C. If the reaction was very slow at 10 °C and too quick to measure at 60 °C, you might narrow the range to 20-50 °C.

4) And they're used to figure out the <u>interval</u> (gaps) between the values too.

<u>Reaction example continued</u>:
If using 10 °C intervals gives you a big change in rate of reaction you might decide to use 5 °C intervals, e.g. 20, 25, 30, 35...

5) Trial runs can also help you figure out <u>how many times</u> the experiment has to be <u>repeated</u> to get reliable results. E.g. if you repeat it two times and the <u>results</u> are all <u>similar</u>, then two repeats is enough.

Collecting Data

Data Should be as *Reliable*, *Accurate* and *Precise* as Possible

Reliable

1) When carrying out an investigation, you can improve the reliability of your results (see p. 3) by repeating the readings and calculating the mean (average, see page 9). You should repeat readings at least twice (so that you have at least three readings to calculate an average result).

2) To make sure your results are reliable you can cross check them by taking a second set of readings with another instrument (or a different observer).

3) Checking your results match with secondary sources, e.g. studies that other people have done, also increases the reliability of your data.

Accurate

1) You should always make sure that your results are accurate. Really accurate results are those that are really close to the true answer.

2) You can get accurate results by doing things like making sure the equipment you're using is sensitive enough (see previous page), and by recording your data to a suitable level of accuracy. For example, if you're taking digital readings of something, the results will be more accurate if you include at least a couple of decimal places instead of rounding to whole numbers.

Precise

1) You should also always make sure your results are precise.

2) Precise results are ones where the data is all really close to the mean (i.e. not spread out).

You Can Check For *Mistakes Made* When *Collecting Data*

1) When you've collected all the results for an experiment, you should have a look to see if there are any results that don't seem to fit in with the rest.

2) Most results vary a bit, but any that are totally different are called anomalous results.

> Barry's results
> measurement 1 — 2.34 cm   measurement 2 — 5.67 cm
> measurement 3 — 2.35 cm   measurement 4 — 2.33 cm

Measurement 2 is an anomalous result — it's totally different from the rest.

3) They're caused by human errors, e.g. by a mistake when measuring.

4) The only way to stop them happening is by taking all your measurements as carefully as possible.

5) If you ever get any anomalous results, you should investigate them to try to work out what happened. If you can work out what happened (e.g. you measured something wrong) you can ignore them when processing your results.

Results need to be reliable, accurate and precise

All this stuff is really important — without good quality data an investigation will be totally meaningless. Get to know this page and your data will be the envy of the whole scientific community.

Organising and Processing Data

The fun doesn't stop once the data's been collected — it then needs to be <u>organised</u> and **processed**...

Data *Needs to be* Organised

1) Data that's been collected needs to be <u>organised</u> so it can be processed later on.

2) <u>Tables</u> are dead useful for <u>organising data</u>.

3) When drawing tables you should always make sure that <u>each column</u> has a <u>heading</u> and that you've included the <u>units</u>.

Test tube	Result (ml)	Repeat 1 (ml)	Repeat 2 (ml)
A	28	37	32
B	47	51	60
C	68	72	70

4) Annoyingly, tables are about as useful as a chocolate teapot for showing <u>patterns</u> or <u>relationships</u> in data. You need to use some kind of graph or mathematical technique for that...

Data *Can be* Processed *Using a Bit of* Maths

1) <u>Raw data</u> generally just ain't that useful. You usually have to <u>process</u> it in some way.

2) A couple of the most simple calculations you can perform are the <u>mean</u> (average) and the <u>range</u> (how spread out the data is).

Mean: *the average*

To calculate the <u>mean</u> <u>ADD TOGETHER</u> all the data values and <u>DIVIDE</u> by the total number of values. You usually do this to get a single value from several <u>repeats</u> of your experiment.

Test tube	Result (ml)	Repeat 1 (ml)	Repeat 2 (ml)	Mean (ml)	Range (ml)
A	28	37	32	(28 + 37 + 32) ÷ 3 = 32.3	37 − 28 = 9
B	47	51	60	(47 + 51 + 60) ÷ 3 = 52.7	60 − 47 = 13
C	68	72	70	(68 + 72 + 70) ÷ 3 = 70.0	72 − 68 = 4

Range: *how spread out the data is*

To calculate the <u>range</u> find the <u>LARGEST</u> number and <u>SUBTRACT</u> the <u>SMALLEST</u> number. You usually do this to <u>check</u> the accuracy and reliability of the results — the <u>greater</u> the <u>spread</u> of the data, the <u>lower</u> the accuracy and reliability.

Processing data requires a teeny bit of maths — don't panic

This data stuff is pretty straightforward — but it's <u>really important</u>. Different measurements scattered on random bits of paper at the bottom of your bag just ain't gonna cut it. Get your ruler out, draw a lovely table, pop your data in, calculate the mean and the range and <u>breathe</u>... Much better.

Presenting Data

Data has to be <u>presented</u>, and graphs are just about the best way of doing it...

Different Types of *Data* Should be *Presented* in *Different Ways*

1) Once you've carried out an investigation, you'll need to <u>present</u> your data so that it's easier to see <u>patterns</u> and <u>relationships</u> in the data.

2) Different types of investigations give you <u>different types</u> of data, so you'll always have to <u>choose</u> what the best way to present your data is.

Pie charts can be used to present the same sort of data as bar charts. They're mostly used when the data is in percentages or fractions though.

Bar Charts

If the independent variable is <u>categoric</u> (comes in distinct categories, e.g. blood types, metals) you should use a <u>bar chart</u> to display the data. You also use them if the independent variable is <u>discrete</u> (the data can be counted in chunks, where there's no in-between value, e.g. number of people is discrete because you can't have half a person).

There are some <u>golden rules</u> you need to follow for <u>drawing</u> bar charts:

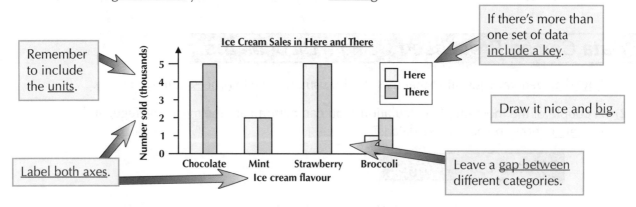

Remember to include the <u>units</u>.

If there's more than one set of data <u>include a key</u>.

Draw it nice and <u>big</u>.

Leave a <u>gap between</u> different categories.

Label both axes.

Line Graphs

If the independent variable is <u>continuous</u> (numerical data that can have any value within a range, e.g. length, volume, temperature) you should use a <u>line graph</u> to display the data.

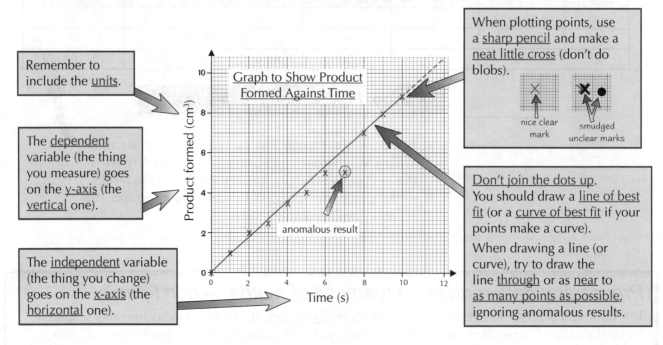

Remember to include the <u>units</u>.

The <u>dependent</u> variable (the thing you measure) goes on the <u>y-axis</u> (the <u>vertical</u> one).

The <u>independent</u> variable (the thing you change) goes on the <u>x-axis</u> (the <u>horizontal</u> one).

When plotting points, use a <u>sharp pencil</u> and make a <u>neat little cross</u> (don't do blobs).

nice clear mark

smudged unclear marks

<u>Don't join the dots up</u>. You should draw a <u>line of best fit</u> (or a <u>curve of best fit</u> if your points make a curve).

When drawing a line (or curve), try to draw the line <u>through</u> or as <u>near</u> to <u>as many points as possible</u>, ignoring anomalous results.

Interpreting Data

Once you've drawn your graph (using all the tips on the previous page) you need to be able to <u>understand</u> what it's <u>telling you</u>. That's where this page comes in handy.

Line Graphs Can Show Relationships in Data

1) Line graphs are great for showing relationships <u>between two variables</u> (just like other graphs).

2) Here are some of the different types of <u>correlation</u> (relationship) shown on line graphs:

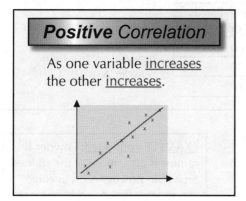

Positive Correlation

As one variable <u>increases</u> the other <u>increases</u>.

Inverse (Negative) Correlation

As one variable <u>increases</u> the other <u>decreases</u>.

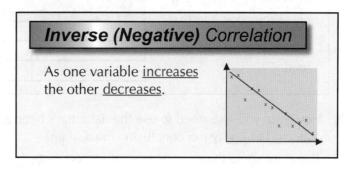

No Correlation

There's <u>no relationship</u> between the two variables.

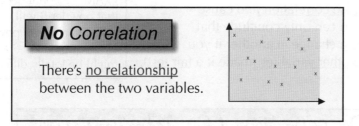

Linear

The graph is a <u>straight line</u>.

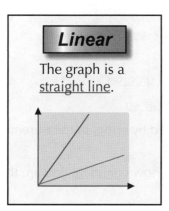

Directly Proportional

The graph is a <u>straight line</u> where both variables increase (or decrease) in the <u>same ratio</u>.

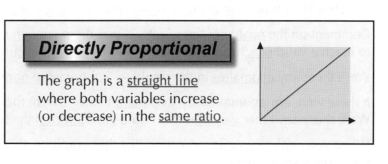

3) You've got to be careful not to <u>confuse correlation</u> with <u>cause</u> though. A <u>correlation</u> just means that there's a <u>relationship</u> between two variables. It <u>doesn't always mean</u> that the change in one variable is <u>causing</u> the change in the other.

4) There are <u>three possible reasons</u> for a correlation. It could be down to <u>chance</u>, it could be that there's a <u>third variable</u> linking the two things, or it might actually be that one variable is <u>causing</u> the other to change.

A correlation is a relationship between sets of data

Wow. What a snazzy page. But don't get distracted by all those crazy lines and crosses — <u>interpreting</u> data's dead important so make sure you know how to do it properly.

Concluding and Evaluating

At the end of an investigation, the <u>conclusion</u> and <u>evaluation</u> are waiting. Don't worry, they won't bite.

A *Conclusion* is a *Summary* of What You've *Learnt*

1) Once all the data's been collected, presented and analysed, an investigation will always involve coming to a <u>conclusion</u>.

2) Drawing a conclusion can be quite straightforward — just <u>look at your data</u> and <u>say what pattern you see</u>.

EXAMPLE: The table on the right shows how effective two washing powders were at removing stains when used at 30 °C.	Washing powder	% of stained areas removed on average	CONCLUSION: Powder <u>B</u> is more effective at removing stains than powder A at <u>30 °C</u>.
	A	60	
	B	80	
	No powder	10	

3) However, you also need to use the data that's been <u>collected</u> to <u>justify</u> the conclusion (back it up).

> EXAMPLE continued: Powder B removed 20% more of the stained area than powder A on average.

4) There are some things to watch out for too — it's important that the conclusion <u>matches the data</u> it's based on and <u>doesn't go any further</u>.

5) Remember not to <u>confuse correlation</u> and <u>cause</u> (see previous page). You can only conclude that one variable is <u>causing</u> a change in another if you have controlled all the <u>other variables</u> (made it a <u>fair test</u>).

> EXAMPLE continued: You can't conclude that powder B will remove more of the stained area than powder A at <u>any other temperature</u> than 30 °C — the results might be totally different.

Evaluations — Describe *How* it Could be *Improved*

An evaluation is a <u>critical analysis</u> of the whole investigation.

1) You should comment on the <u>method</u> — was the <u>equipment suitable</u>? Was it a <u>fair test</u>?

2) Comment on the <u>quality</u> of the <u>results</u> — was there <u>enough evidence</u> to reach a valid <u>conclusion</u>? Were the results <u>reliable</u>, <u>accurate</u> and <u>precise</u>?

3) Were there any <u>anomalies</u> in the results — if there were <u>none</u> then <u>say so</u>.

4) If there were any anomalies, try to <u>explain</u> them — were they caused by <u>errors</u> in measurement? Were there any other <u>variables</u> that could have <u>affected</u> the results?

5) When you analyse your investigation like this, you'll be able to say how <u>confident</u> you are that your conclusion is <u>right</u>.

6) Then you can suggest any <u>changes</u> that would <u>improve</u> the quality of the results, so that you could have <u>more confidence</u> in your conclusion. For example, you might suggest changing the way you controlled a variable, or changing the interval of values you measured.

7) You could also make more <u>predictions</u> based on your conclusion, then <u>further experiments</u> could be carried out to test them.

8) When suggesting improvements to the investigation, always make sure that you say <u>why</u> you think this would make the results <u>better</u>.

An experiment must have a conclusion and an evaluation

I know it doesn't seem very nice, but writing about where you went <u>wrong</u> is an important skill — it shows you've got a really good understanding of what the investigation was <u>about</u>.

The History of the Atom

If you cut up a cake you end up with slices. If you keep going you're gonna have crumbs. But what happens if you keep cutting... Just how small can you go and what would the stuff you end up with look like... Scientists have been trying to work it out for years...

The Theory of **Atomic Structure**
Has **Changed Throughout History**

Atoms are the tiny particles of matter (stuff that has a mass) which make up everything in the universe...

1) At the start of the 19th century John Dalton described atoms as solid spheres, and said that different spheres made up the different elements.

2) In 1897 J J Thomson concluded from his experiments that atoms weren't solid spheres. His measurements of charge and mass showed that an atom must contain even smaller, negatively charged particles — electrons. The 'solid sphere' idea of atomic structure had to be changed. The new theory was known as the 'plum pudding model'.

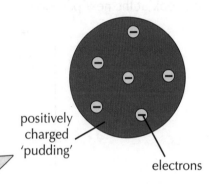

positively charged 'pudding'

electrons

Rutherford Showed that the **Plum Pudding Model** Was Wrong

1) In 1909 Ernest Rutherford and his students Hans Geiger and Ernest Marsden conducted the famous gold foil experiment. They fired positively charged particles at an extremely thin sheet of gold.

2) From the plum pudding model, they were expecting most of the particles to be deflected by the positive 'pudding' that made up most of an atom. In fact, most of the particles passed straight through the gold atoms, and a very small number were deflected backwards. So the plum pudding model couldn't be right.

3) So Rutherford came up with an idea that could explain this new evidence — the theory of the nuclear atom. In this, there's a tiny, positively charged nucleus at the centre, surrounded by a 'cloud' of negative electrons — most of the atom is empty space.

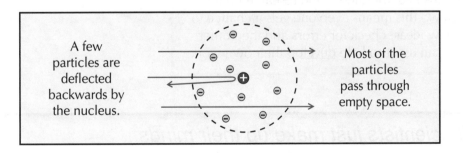

A few particles are deflected backwards by the nucleus.

Most of the particles pass through empty space.

The History of the Atom

The Refined *Bohr Model* Explains a Lot...

1) Scientists realised that electrons in a 'cloud' around the nucleus of an atom, as Rutherford described, would be attracted to the nucleus, causing the atom to <u>collapse</u>. Niels Bohr proposed a new model of the atom where all the electrons were contained in <u>shells</u>.

2) Bohr suggested that electrons can only exist in <u>fixed orbits</u>, or <u>shells</u>, and not anywhere in between. Each shell has a <u>fixed energy</u>.

3) Bohr's theory of atomic structure was supported by many <u>experiments</u> and it helped to explain lots of other scientists' <u>observations</u> at the time. It was <u>pretty close</u> to our currently accepted version of the atom (have a look at the next page to see what we now think atoms look like).

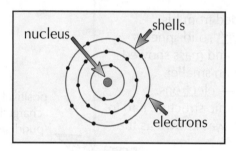

Scientific *Theories* Have to be Backed Up by *Evidence*

1) So, you can see that what we think the atom looks like <u>now</u> is <u>completely</u> <u>different</u> to what people thought in the past. These different ideas were <u>accepted</u> because they fitted the <u>evidence</u> available at the time.

2) As scientists did more <u>experiments</u>, new evidence was found and our theory of the <u>structure</u> of the atom was <u>modified</u> to fit it.

3) This is nearly always the way <u>scientific knowledge</u> develops — new evidence prompts people to come up with new, <u>improved</u> <u>ideas</u>. These ideas can be used to make <u>predictions</u> which if proved correct are a pretty good indication that the ideas are <u>right</u>.

4) Scientists also put their ideas and research up for <u>peer review</u>. This means everyone gets a chance to see the new ideas, check for errors and then other scientists can use it to help <u>develop</u> their own work.

Why can't scientists just make up their minds...

The history of the atom has gone through many stages thanks to people making predictions, finding evidence, changing their minds and making new predictions. A fine example of science in action.

Atoms and Elements

Atoms are <u>amazingly tiny</u> — you can only see them with an incredibly powerful microscope.

Atoms have a Small **Nucleus** Surrounded by **Electrons**

There are quite a few different (and equally useful) models of the atom — but chemists tend to like this <u>model</u> best. You can use it to explain pretty much the whole of Chemistry... which is nice.

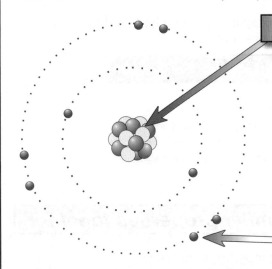

The Nucleus

1) It's in the <u>middle</u> of the atom.
2) It contains <u>protons</u> and <u>neutrons</u>.
3) <u>Protons</u> are <u>positively charged</u>.
4) <u>Neutrons</u> have <u>no charge</u> (they're neutral).
5) So the nucleus has a <u>positive charge</u> overall because of the protons.
6) But size-wise it's <u>tiny</u> compared to the rest of the atom.

The Electrons

1) Move <u>around</u> the nucleus.
2) They're <u>negatively charged</u>.
3) They're <u>tiny</u>, but they cover <u>a lot of space</u>.
4) They occupy <u>shells</u> around the nucleus.
5) These shells explain <u>the whole of Chemistry</u>.

Particle	Relative mass	Relative charge
Proton	1	+1
Neutron	1	0
Electron	$\frac{1}{2000}$	-1

Number of Protons **Equals** Number of Electrons

1) Atoms have <u>no charge</u> overall. They are neutral.
2) The <u>charge</u> on the electrons is the <u>same</u> size as the charge on the <u>protons</u> — but <u>opposite</u>.
3) This means the <u>number</u> of <u>protons</u> always equals the <u>number</u> of <u>electrons</u> in an <u>atom</u>.
4) If some electrons are <u>added or removed</u>, the atom becomes <u>charged</u> and is then an <u>ion</u>.

Elements Consist of **One Type** of Atom Only

1) Atoms can have different numbers of protons, neutrons and electrons. It's the number of <u>protons</u> in the nucleus that decides what <u>type</u> of atom it is.
2) For example, an atom with <u>one proton</u> in its nucleus is <u>hydrogen</u> and an atom with <u>two protons</u> is <u>helium</u>.
3) If a substance only contains <u>one type</u> of atom it's called an <u>element</u>.
4) There are about <u>100 different elements</u> — quite a lot of everyday substances are elements:

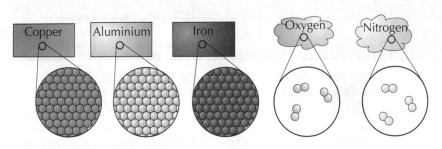

So <u>all the atoms</u> of a particular <u>element</u> (e.g. nitrogen) have the <u>same number</u> of protons...

...and <u>different elements</u> have atoms with <u>different</u> <u>numbers</u> of protons.

The Periodic Table

Chemistry would be really messy if it was all big lists of names and properties. So instead they've come up with a kind of shorthand for the names, and made a beautiful table to organise the elements — like a big filing system.

Atoms can be represented by symbols

Atoms of each element can be represented by a one or two letter symbol — it's a type of shorthand that saves you the bother of having to write the full name of the element.

Some make perfect sense, e.g.

> **C = carbon**   **Li = lithium**   **Mg = magnesium**

Others seem to make about as much sense as an apple with a handle.

E.g.

> **Na = sodium**   **Fe = iron**   **Pb = lead**

Most of these odd symbols actually come from the Latin names of the elements.

The periodic table puts elements with similar properties together

1) The periodic table is laid out so that elements with similar properties form columns.

2) These vertical columns are called groups and Roman numerals are often used for them.

3) All of the elements in a group have the same number of electrons in their outer shell.

4) This is why elements in the same group have similar properties. So, if you know the properties of one element, you can predict properties of other elements in that group.

5) For example, the Group 1 elements are Li, Na, K, Rb, Cs and Fr. They're all metals and they react the same way. E.g. they all react with water to form an alkaline solution and hydrogen gas, and they all react with oxygen to form an oxide.

6) The elements in the final column (Group 0) are the noble gases. They all have eight electrons in their outer shell, apart from helium (which has two). This means that they're stable and unreactive.

The top number is the mass number. This is the total number of protons and neutrons.

So, if you want to find the number of neutrons in an atom, just subtract the atomic number from the mass number.

The bottom number is the atomic number. This is the number of protons, which conveniently also tells you the number of electrons.

Atomic number is also known as proton number.

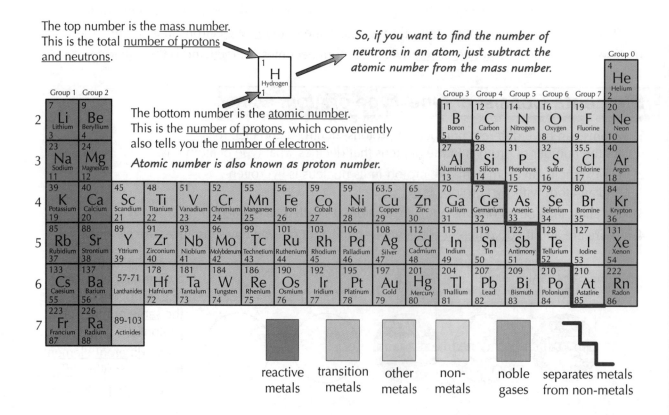

Electron Shells

The fact that electrons occupy "shells" around the nucleus is what causes the whole of chemistry. Remember that, and watch how it applies to each bit of it. It's ace.

Electron shell **rules:**

1) Electrons always occupy <u>shells</u> (sometimes called <u>energy levels</u>).

2) The <u>lowest</u> energy levels are <u>always filled first</u> — these are the ones closest to the nucleus.

3) Only <u>a certain number</u> of electrons are allowed in each shell:

- <u>1st shell</u> — 2
- <u>2nd shell</u> — 8
- <u>3rd shell</u> — 8

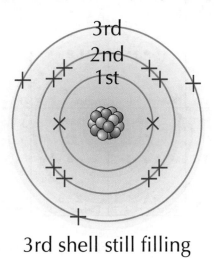

3rd shell still filling

4) Atoms are much <u>happier</u> when they have <u>full electron shells</u>
   — like the <u>noble gases</u> in <u>Group 0</u>.

5) In most atoms the <u>outer shell</u> is <u>not full</u> and this makes the atom want to <u>react</u> to fill it.

Electron shells — probably the most important thing in chemistry

It's really important to learn the rules for filling electron shells. It's so important I'll just leave this page with a quick reminder. Energy of the shells increases with increasing number (so shell one is the lowest). Fill the shell with lowest energy first. And the 1st shell can only hold a maximum of 2 electrons, but the 2nd and 3rd shells can both hold 8 electrons. Practise following these rules on the next page.

Electron Shells

You can use the electron shell rules from the previous page to work out the <u>electronic structures</u> for the first <u>20</u> elements.

*Follow the rules to **work out** electronic structures*

Electronic structures are not hard to work out.
For a quick example, take nitrogen. <u>Follow the steps...</u>

> 1) The periodic table tells us nitrogen has <u>seven</u> protons... so it must have <u>seven</u> electrons.
>
> 2) Follow the '<u>Electron Shell Rules</u>' from the previous page. The <u>first</u> shell can only take 2 electrons and the <u>second</u> shell can take a <u>maximum</u> of 8 electrons.
>
> 3) So the electronic structure for nitrogen <u>must</u> be <u>2, 5</u>. Easy peasy.

Now <u>you</u> try it for argon.

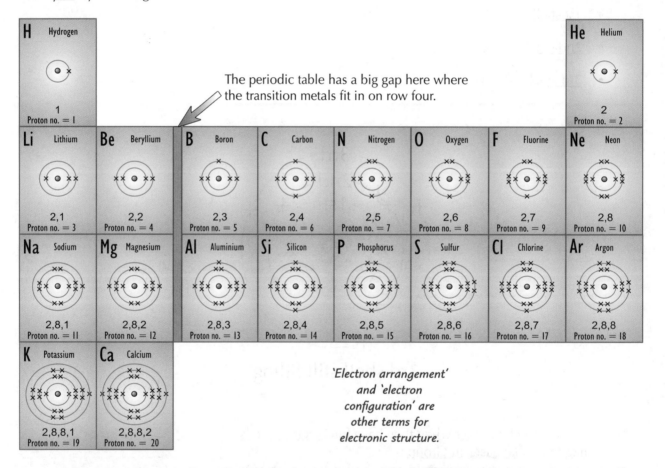

The periodic table has a big gap here where the transition metals fit in on row four.

*'Electron arrangement'
and 'electron
configuration' are
other terms for
electronic structure.*

> <u>Answer...</u>
>
> To calculate the electronic structure of argon, <u>follow the rules</u>. It's got 18 protons, so it <u>must</u> have 18 electrons. The first shell must have <u>2</u> electrons, the second shell must have <u>8</u>, and so the third shell must have <u>8</u> as well. It's as easy as <u>2, 8, 8</u>.

Each shell can only take a set number shof electrons

The electron shell rules are great — they mean you don't have to learn each element separately.
If you <u>learn the pattern</u> you'll be able to work out the electronic structure of <u>any</u> of the first 20 elements.
Cover the page: using a periodic table, find the atom with the electronic structure 2, 8, 6.*

Compounds

Life'd be oh so simple if you only had to worry about elements, even if there are a hundred or so of them. But you can mix and match elements to make lots of compounds, which complicates things no end.

Atoms *join together* to make *compounds*

1) When <u>different elements react</u>, atoms form <u>chemical bonds</u> with other atoms to form <u>compounds</u>. It's <u>usually difficult</u> to <u>separate</u> the two original elements out again.

2) <u>Making bonds</u> involves atoms giving away, taking or sharing <u>electrons</u>. Only the <u>electrons</u> are involved — it's nothing to do with the nuclei of the atoms at all.

3) A compound which is formed from a <u>metal</u> and a <u>non-metal</u> consists of <u>ions</u>. The <u>metal</u> atoms <u>lose</u> electrons to form <u>positive ions</u> and the non-metal atoms <u>gain</u> electrons to form <u>negative ions</u>. The <u>opposite charges</u> (positive and negative) of the ions mean that they're strongly <u>attracted</u> to each other. This is called <u>IONIC bonding</u>.

   E.g. NaCl

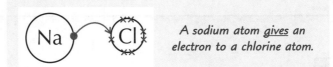

A sodium atom gives an electron to a chlorine atom.

4) A compound formed from <u>non-metals</u> consists of <u>molecules</u>. Each atom <u>shares</u> an <u>electron</u> with another atom — this is called a <u>COVALENT bond</u>. Each atom has to make enough covalent bonds to <u>fill up</u> its <u>outer shell</u>.

   E.g. HCl

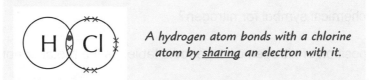

A hydrogen atom bonds with a chlorine atom by <u>sharing</u> an electron with it.

5) The <u>properties</u> of a compound are <u>totally different</u> from the properties of the <u>original elements</u>. For example, if iron (a lustrous magnetic metal) and sulfur (a nice yellow powder) react, the compound formed (<u>iron sulfide</u>) is a <u>dull grey solid lump</u>, and doesn't behave <u>anything like</u> either iron or sulfur.

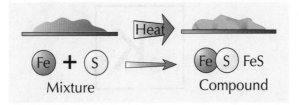

6) Compounds can be <u>small molecules</u> like water, or <u>great whopping lattices</u> like sodium chloride (when I say whopping I'm talking in atomic terms).

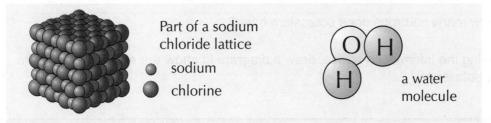

Part of a sodium chloride lattice

- sodium
- chlorine

a water molecule

Warm-Up and Exam Questions

It's easy to think you've learnt everything in the section until you try the warm-up questions. Don't panic if there are bits you've forgotten. Just go back over those bits until they're firmly fixed in your brain.

Warm-Up Questions

1) Describe J J Thomson's model of the atom.
2) In a neutral atom, which particles are always equal in number?
3) What is the definition of an element? Roughly how many different elements are there?
4) Explain the difference between mass number and atomic number.
5) How many electrons can be held in:
   a) the first shell, and
   b) the second shell?
6) Aluminium has 13 protons. Work out the electronic structure of aluminium.
7) What does a compound formed from non-metals consist of?

Exam Questions

1 Nitrogen has 7 protons and 7 neutrons.

   (a) How many electrons does nitrogen have?

   (1 mark)

   (b) What is the chemical symbol for nitrogen?

   (1 mark)

   (c) Look at the position of nitrogen on a periodic table. Is it a metal or non-metal?

   (1 mark)

   (d) Give another element that will have similar chemical properties to nitrogen. Explain why you chose this element.

   (3 marks)

2 The diagram below shows potassium as it appears in the periodic table.

$$^{39}_{19}K$$

   (a) How many electrons does potassium have?

   (1 mark)

   (b) How many protons does potassium have?

   (1 mark)

   (c) How many neutrons does potassium have?

   (1 mark)

   (d) Using the information above, draw a diagram to show the electronic structure of potassium.

   (2 marks)

Exam Questions

3   The electron arrangement of sodium is shown in the diagram:

(a)   How many protons does sodium have?

(1 mark)

(b)   Sodium is in Group 1. Give another element that would have the same
       number of outer shell electrons.

(1 mark)

(c)   How many electrons does sodium need to lose so that it has a full outer shell?

(1 mark)

4   A compound formed from a metal and a non-metal will consist of ions.

(a)   Explain how the ions are formed and why they are strongly attracted to each other.

(3 marks)

(b)   Name and describe the type of bonding that occurs between
       the atoms in a compound formed from non-metals.

(2 marks)

5   The proton has a relative mass of 1. What is the relative mass of the neutron?

A   2000

B   1

C   1/2000

D   2

(1 mark)

6   Which of the following statements about the groups of the periodic table is not true?

A   A group is one vertical column of the periodic table.

B   The elements in a group have the same number of electrons in their outer shells.

C   All of the elements in a group have different chemical properties.

D   The elements in Group 0 of the periodic table are the noble gases

(1 mark)

7   *In this question you will be assessed on the quality of your English, the organisation of your
     ideas and your use of appropriate specialist vocabulary.*

Describe how the theory of atomic structure has changed throughout history.

(6 marks)

Formulas and Reactions

Every compound has a formula — it tells you what it's made up of.

Formulas Show the Atoms in a Substance

You can work out how many atoms of each type there are in a substance when you're given its formula. Here are some examples:

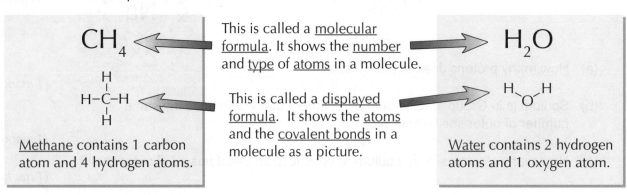

CH_4

This is called a molecular formula. It shows the number and type of atoms in a molecule.

H_2O

H–C–H with H above and H below

This is called a displayed formula. It shows the atoms and the covalent bonds in a molecule as a picture.

H–O–H

Methane contains 1 carbon atom and 4 hydrogen atoms.

Water contains 2 hydrogen atoms and 1 oxygen atom.

Don't panic if the molecular formula has brackets in it. They're easy too.

$CH_3(CH_2)_2CH_3$

The 2 after the bracket means that there are 2 lots of CH_2. So altogether there are 4 carbon atoms and 10 hydrogen atoms.

Drawing the displayed formula of the compound is a good way to count up the number of atoms.

Do it a bit at a time.

$CH_3(CH_2)_2CH_3$

H H H H
H-C-C-C-C-H
H H H H

Atoms Aren't Lost or Made in Chemical Reactions

1) During chemical reactions, things don't appear out of nowhere and things don't just disappear.

2) You still have the same atoms at the end of a chemical reaction as you had at the start. They're just arranged in different ways.

How to balance symbol equations is covered on page 23.

3) Balanced symbol equations show the atoms at the start (the reactant atoms) and the atoms at the end (the product atoms) and how they're arranged. For example:

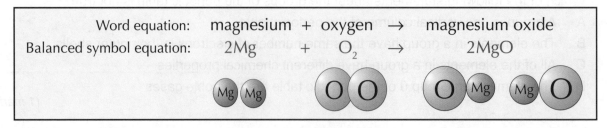

Word equation: magnesium + oxygen → magnesium oxide
Balanced symbol equation: $2Mg$ + O_2 → $2MgO$

4) Because atoms aren't gained or lost, the mass of the reactants equals the mass of the products. So, if you completely react 6 g of magnesium with 4 g of oxygen, you'd end up with 10 g of magnesium oxide.

Not learning this stuff will only compound your problems...

Formulas are pretty useful. The formula of a compound tells you what atoms it's made up of. You can use formulas to write out balanced symbol equations (see next page) for reactions. And remember — you'll always have the same number of atoms at the end of a reaction as at the start.

Balancing Equations

Equations need a lot of practice if you're going to get them right — don't just skate over this stuff.

Balancing the Equation — Match Them Up One by One

1) There must always be the same number of atoms of each element on both sides —
   they can't just disappear.

2) You balance the equation by putting numbers in front of the formulas where needed.
   Take this equation for reacting sulfuric acid (H_2SO_4) with sodium hydroxide ($NaOH$)
   to get sodium sulfate (Na_2SO_4) and water (H_2O):

$$H_2SO_4 \ + \ NaOH \ \rightarrow \ Na_2SO_4 \ + \ H_2O$$

The formulas are all correct but the numbers of some atoms don't match up on both sides.
E.g. there are 3 Hs on the left, but only 2 on the right. You can't change formulas like H_2O to H_3O.
You can only put numbers in front of them.

Method: Balance Just ONE Type of Atom at a Time

The more you practise, the quicker you get, but all you do is this:

1) Find an element that doesn't balance and pencil in a number to try and sort it out.
2) See where it gets you. It may create another imbalance — if so, just pencil in
   another number and see where that gets you.
3) Carry on chasing unbalanced elements and it'll sort itself out pretty quickly.

I'll show you. In the equation above you soon notice we're short of H atoms on the
RHS (Right-Hand Side).

1) The only thing you can do about that is make it $2H_2O$ instead of just H_2O:

$$H_2SO_4 \ + \ NaOH \ \rightarrow \ Na_2SO_4 \ + 2H_2O$$

2) But that now causes too many H atoms and O atoms on the RHS, so to balance that up you
   could try putting $2NaOH$ on the LHS (Left-Hand Side):

$$H_2SO_4 \ + \ 2NaOH \ \rightarrow \ Na_2SO_4 \ + 2H_2O$$

3) And suddenly there it is! Everything balances. And you'll notice the Na just sorted itself out.

Balancing equations — weigh it up in your mind...

REMEMBER WHAT THOSE NUMBERS MEAN: A number in front of a formula applies to the
entire formula. So, $3Na_2SO_4$ means three lots of Na_2SO_4. The little numbers in the middle or at
the end of a formula only apply to the atom or brackets immediately before.
So the 4 in Na_2SO_4 just means 4 Os, not 4 Ss.

Kinetic Theory and Forces Between Particles

Heating things up often causes them to change state. This is because of the forces involved...

States of Matter — Depend on the Forces Between Particles

All stuff is made of particles (molecules, ions or atoms) that are constantly moving, and the forces between these particles can be weak or strong, depending on whether it's a solid, liquid or a gas.

Solids

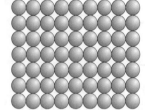

1) There are strong forces of attraction between particles, which holds them in fixed positions in a very regular lattice arrangement.

2) The particles don't move from their positions, so all solids keep a definite shape and volume, and don't flow like liquids.

3) The particles vibrate about their positions — the hotter the solid becomes, the more they vibrate (causing solids to expand slightly when heated).

If you heat the solid (give the particles more energy), eventually the solid will melt and become liquid.

Liquids

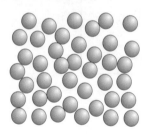

1) There is some force of attraction between the particles. They're free to move past each other, but they do tend to stick together.

2) Liquids don't keep a definite shape and will flow to fill the bottom of a container.

3) The particles are constantly moving with random motion. The hotter the liquid gets, the faster they move. This causes liquids to expand slightly when heated.

If you now heat the liquid, eventually it will boil and become gas.

Gases

1) There's next to no force of attraction between the particles — they're free to move. They travel in straight lines and only interact when they collide.

2) Gases don't keep a definite shape or volume and will always fill any container. When particles bounce off the walls of a container they exert a pressure on the walls.

3) The particles move constantly with random motion. The hotter the gas gets, the faster they move. Gases either expand when heated, or their pressure increases.

Volatility and Solutions

You can explain a lot of things (including <u>perfumes</u>) if you get your head round this lot.

*How We Smell Stuff — **Volatility's** the Key*

1) When a <u>liquid</u> is <u>heated</u>, the heat energy goes to the particles, which makes them <u>move faster</u>.

2) <u>Some</u> particles move <u>faster</u> than others.

3) Fast-moving particles <u>at the surface</u> will <u>overcome</u> the <u>forces of attraction</u> from the other particles and <u>escape</u>. This is <u>evaporation</u>.

4) How <u>easily</u> a liquid evaporates is called its <u>volatility</u>.

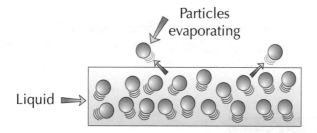

Particles evaporating

Liquid

So... the evaporated particles are now drifting about in the air, the <u>smell receptors</u> in your <u>nose</u> pick up the chemical — and hey presto — you <u>smell</u> it.

Perfumes need to be <u>quite volatile</u> so that they can evaporate enough for you to smell them. The particles in liquid perfumes only have a very <u>weak attraction</u> between them. It's easy for the particles to overcome this and <u>escape</u> — so you only need a very little heat energy to make the perfume evaporate.

*A **Solution** is a **Mixture** of **Solvent** and **Solute***

1) When you add a <u>solid</u> (the <u>solute</u>) to a <u>liquid</u> (the <u>solvent</u>) the bonds holding the solute molecules together <u>sometimes break</u> and the molecules then <u>mix</u> with the molecules in the liquid — forming a <u>solution</u>. This is called <u>dissolving</u>.

2) Whether the bonds break depends on how <u>strong</u> the <u>attractions</u> are between the molecules <u>within</u> each substance and how strong the attractions are <u>between</u> the two substances.

Here are some important <u>definitions</u>:

1) <u>Solution</u> – is a mixture of a solute and a solvent that does not separate out.

2) <u>Solute</u> – is the substance being dissolved.

3) <u>Solvent</u> – is the liquid it's dissolving into.

4) <u>Soluble</u> – means it will dissolve.

5) <u>Insoluble</u> – means it will <u>NOT</u> dissolve.

6) <u>Solubility</u> – a measure of how much will dissolve.

E.g. <u>brine</u> is a solution of <u>salt</u> and <u>water</u> — if you <u>evaporated</u> off the <u>solvent</u> (the water), you'd see the <u>solute</u> (the salt) again.

Water is a very common solvent.

Solutions

Solutions are all around you — e.g. sea water, bath salts... And inside you even — e.g. instant coffee...

Nail Varnish is Insoluble in Water...

Nail varnish doesn't dissolve in water. This is for two reasons:

1) The molecules of nail varnish are strongly attracted to each other. This attraction is stronger than the attraction between the nail varnish molecules and the water molecules.

2) The molecules of water are strongly attracted to each other. This attraction is stronger than the attraction between the water molecules and the nail varnish molecules.

> Because the two substances are more attracted to themselves than each other, they don't form a solution.

...but Soluble in Acetone

Nail varnish dissolves in acetone — more commonly known as nail varnish remover. This is because the attraction between acetone molecules and nail varnish molecules is stronger than the attractions holding the two substances together.

So the solubility of a substance depends on the solvent used.

Acetone is also called propanone.

Lots of Things are Solvents

Alcohols and esters can be used as solvents, and so can lots of other weird and wacky organic molecules. Ability to dissolve a solute isn't the only consideration though... some solvents are horribly poisonous.

Example: Mothballs are made of a substance called naphthalene. Imagine you've trodden a mothball into your carpet. Choose one of the solvents from the table to clean it up.

Solvent	Solubility of naphthalene	Boiling point	Other properties
water	0 g/100 g	100 °C	safe
methanol	9.7 g/100 g	65 °C	flammable
ethyl acetate	18.5 g/100 g	77 °C	flammable
dichloromethane	25.0 g/100 g	40 °C	extremely toxic

Looking at the data, water wouldn't be a good choice because it doesn't dissolve the naphthalene. Dichloromethane would dissolve it easily, but it's very toxic. Of the two solvents left, ethyl acetate dissolves more naphthalene (so you won't need as much). Ethyl acetate is best (just don't set light to it).

This page has all of the solutions...

If you ever spill bright pink nail varnish (or any other colour for that matter) on your carpet, go easy with the nail varnish remover. If you use too much, the nail varnish dissolves in the remover, forming a solution which can go everywhere — and you end up with an enormous bright pink stain... aaagh.

Warm-Up and Exam Questions

You've almost reached the end of this section, but don't stop just yet. Take a look at these questions to make sure you understand everything you've just read.

Warm-Up Questions

1) Draw the displayed formula for water.

2) Balance this equation for the reaction of glucose ($C_6H_{12}O_6$) and oxygen:
$$C_6H_{12}O_6 + O_2 \rightarrow CO_2 + H_2O$$

3) Look at this equation:
$$H_2SO_4 + 2NaOH \rightarrow Na_2SO_4 + 2H_2O$$
Explain the difference between the meaning of the $_2$ in H_2SO_4 and the 2 in 2NaOH.

4) What happens to a gas when it is heated?

5) A solution is made up of two parts. What are they?

Exam Questions

1 (a) Using the ideas of kinetic theory, explain:

(i) why liquids flow.

(2 marks)

(ii) why gases fill their containers.

(2 marks)

(iii) why solids have a fixed shape.

(2 marks)

(b) (i) Explain the process of evaporation, in terms of particles.

(1 mark)

(ii) Explain why perfumes are made from substances which evaporate easily.

(1 mark)

2 Which of these statements about chemical reactions is **not** true?

A The mass of the reactants is always equal to the mass of the products.

B Atoms are neither created nor destroyed in a reaction.

C The mass of the products is always less than the mass of the reactants.

D In a written equation, the mass of all the atoms on the left of the arrow is equal to the mass of all the atoms on the right of the arrow.

(1 mark)

Exam Questions

3   A strong glue contains cyanoacrylate which causes objects to stick to each other.
    The table shows some possible solvents which could be used to separate two objects
    that have been joined using the glue.

Solvent	Solubility of cyanoacrylate	Boiling point	Other properties
water	0 g/100 g	100 °C	safe
acetone	16 g/100 g	57 °C	flammable
ethanol	5.8 g/100 g	79 °C	flammable

(a) Cyanoacrylate is a solute. Explain what this means.

(1 mark)

(b) Look at the table. Explain why you would use acetone to separate two objects
     that have been joined using the glue, rather than water.

(2 marks)

4   Sulfuric acid, H_2SO_4, reacts with ammonia, NH_3, to form ammonium sulfate, $(NH_4)_2SO_4$.

(a) Write the word equation for this reaction.

(1 mark)

(b) Write a balanced symbol equation for this reaction.

(2 marks)

(c) In the balanced equation, how many atoms are there in the reactants?

(1 mark)

5   The equation for a reaction is shown below.

$$X + Y \rightarrow Z$$

Substance **Y** reacts with 4 g of substance **X**. 17 g of substance **Z** is produced.

(a) Describe what happens to the number of atoms during a chemical reaction.

(1 mark)

(b) Work out the mass of substance **Y** involved in the reaction.

(2 marks)

(c) Substance **Z** is insoluble in water.

(i)   Describe what happens when a substance dissolves in water.

(1 mark)

(ii) Suggest two reasons why substance **Z** does not dissolve in water.

(2 marks)

Revision Summary for Section 1

There wasn't anything too ghastly in this section, and a few bits were even quite interesting I reckon. But you've got to make sure the facts are all firmly embedded in your brain and that you really understand what's going on. These questions will let you see what you know and what you don't. If you get stuck on any, you need to look at that stuff again. Keep going till you can do them all without coming up for air.

1) Describe the famous 'gold foil experiment'. What did Rutherford conclude from it?

2) Sketch an atom. Label the nucleus and the electrons.

3) State the relative mass and charge of each particle in an atom.

4) What are the symbols for:
a) calcium?
b) carbon?
c) sodium?

5)* Which element's properties are more similar to magnesium's: calcium or iron?

6) The element boron is written as $^{11}_{5}B$. How many neutrons does an atom of this element contain? How many electrons does a neutral boron atom have?

7) Describe how you would work out the electronic structure of an atom given its atomic number.

8) Calculate the electronic structures for each of the following elements: $^{4}_{2}He$, $^{12}_{6}C$, $^{31}_{15}P$.

9) Draw diagrams to show the electronic structures for the first 20 elements.

10) Describe the process of ionic bonding.

11) What is covalent bonding?

12)* A molecule has the molecular formula $CH_3(CH_2)_4CH_3$. How many H and C atoms does it contain?

13)* Write down the displayed formula for a molecule with the molecular formula C_3H_8.

14)* Balance these equations:
a) $CaCO_3 + HCl \rightarrow CaCl_2 + H_2O + CO_2$
b) $Ca + H_2O \rightarrow Ca(OH)_2 + H_2$

15) A substance keeps the same volume, but changes its shape according to the container it's held in. Is it a solid, a liquid or a gas? How strong are the forces of attraction between its particles?

16) What does it mean if a liquid is said to be very volatile?

17) In salt water, what is:
a) the solute?
b) the solution?

18) Explain why nail varnish doesn't dissolve in water.

* Answers on page 251

Natural and Synthetic Materials

<u>Materials</u> are just the substances that we use to make stuff — they can be <u>natural</u> or <u>man-made</u>.

All Materials are Made Up of Chemicals

1) We use a wide range of materials, including <u>metals</u>, <u>ceramics</u> and <u>polymers</u>. Absolutely every material is made up of <u>chemicals</u>, either <u>individual chemicals</u> or <u>mixtures</u> of chemicals.

2) Chemicals are made up of <u>atoms</u> or <u>groups of atoms</u> bonded together called compounds (see p 19).

3) Some materials are <u>mixtures</u> of chemicals. A mixture contains different substances that are <u>not</u> chemically bonded together. For example, <u>rock salt</u> is a mixture of two compounds — salt and sand.

Some of These Materials Occur Naturally...

A lot of the materials we use come from other <u>living things</u>:

MATERIAL FROM PLANTS	MATERIAL FROM ANIMALS
1) <u>Wood</u> and <u>paper</u> are both made from <u>trees</u>. 2) <u>Cotton</u> comes from the cotton plant.	1) <u>Wool</u> comes from <u>sheep</u>. 2) <u>Silk</u> is made by the <u>silkworm</u> larva. 3) <u>Leather</u> comes from <u>cows</u>.

...Others are Synthetic — Made by Humans

We also use <u>man-made</u> (synthetic) materials:

1) All <u>rubber</u> used to come from the sap of the <u>rubber tree</u>. We <u>still</u> get a lot of rubber this way (e.g. for car tyres), but you can also make rubber in a <u>factory</u>. The advantage of this is that you can <u>control</u> its <u>properties</u>, making it suitable for different <u>purposes</u>, e.g. wetsuits.

2) A lot of <u>clothes</u> are made of <u>man-made</u> fabrics like <u>nylon</u> or <u>polyester</u>. As with synthetic rubbers, the properties of synthetic fabrics can be <u>controlled</u> by the manufacturer — e.g. you can make fabrics that are <u>water-proof</u>, <u>super-stretchy</u>, or <u>sparkly</u>.

3) Most <u>paints</u> are mixtures of man-made chemicals. The <u>pigment</u> (the colouring) and the stuff that holds it all together are designed to be <u>tough</u> and to stop the colour fading.

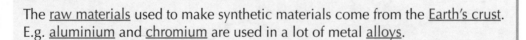

The <u>raw materials</u> used to make synthetic materials come from the <u>Earth's crust</u>. E.g. <u>aluminium</u> and <u>chromium</u> are used in a lot of metal <u>alloys</u>.

So silk comes from worms...

Okay, so now you've read this page, you should know what a <u>material</u> is (and that it doesn't just mean <u>fabric</u>). You should also have a pretty good idea of the different kinds of materials around, and where they come from. Not bad for the first page of the section, I think you'll agree.

Using Limestone

Limestone's often formed from <u>sea shells</u>, so you might not expect it to be useful as a <u>building material</u>.

Limestone is Mainly Calcium Carbonate

St Paul's Cathedral is made from limestone.

Limestone's <u>quarried</u> out of the ground — it's great for making into <u>blocks</u> for building with. Fine old buildings like <u>cathedrals</u> are often made purely from limestone blocks. It's pretty <u>sturdy</u> stuff, but don't go thinking it doesn't <u>react</u> with anything.

1) Limestone is mainly <u>calcium carbonate</u> — $CaCO_3$.

2) When it's heated it <u>thermally decomposes</u> to make <u>calcium oxide</u> and <u>carbon dioxide</u>.

> **calcium carbonate → calcium oxide + carbon dioxide**
> $$CaCO_{3(s)} \rightarrow CaO_{(s)} + CO_{2(g)}$$

Thermal decomposition is when one substance chemically changes into at least two new substances when it's heated.

- When <u>magnesium</u>, <u>copper</u>, <u>zinc</u> and <u>sodium carbonates</u> are heated, they decompose in the <u>same way</u>.
  E.g. magnesium carbonate → magnesium oxide + carbon dioxide (i.e. $MgCO_3 \rightarrow MgO + CO_2$)
- However, you might have <u>difficulty</u> doing some of these reactions in class — a <u>Bunsen burner</u> can't reach a <u>high enough temperature</u> to thermally decompose some carbonates of <u>Group I metals</u>.

3) Calcium carbonate also reacts with <u>acid</u> to make a <u>calcium salt</u>, <u>carbon dioxide</u> and <u>water</u>. E.g.:

> **calcium carbonate + sulfuric acid → calcium sulfate + carbon dioxide + water**
> $$CaCO_3 + H_2SO_4 \rightarrow CaSO_4 + CO_2 + H_2O$$

- The type of <u>salt</u> produced <u>depends</u> on the type of <u>acid</u>. For example, a reaction with <u>hydrochloric</u> acid would make a <u>chloride</u> (e.g. $CaCl_2$).
- Other carbonates that react with acids are <u>magnesium</u>, <u>copper</u>, <u>zinc</u> and <u>sodium</u>.

This reaction means that limestone is damaged by acid rain (see p. 91).

Calcium Oxide Reacts with Water to Produce Calcium Hydroxide

1) When you <u>add water</u> to calcium oxide you get <u>calcium hydroxide</u>.

> **calcium oxide + water → calcium hydroxide**

> or $CaO + H_2O \rightarrow Ca(OH)_2$

2) Calcium hydroxide is an <u>alkali</u> which can be used to neutralise <u>acidic soil</u> in fields. Powdered limestone can be used for this too, but the <u>advantage</u> of <u>calcium hydroxide</u> is that it works <u>much faster</u>.

3) Calcium hydroxide can also be used in a <u>test</u> for <u>carbon dioxide</u>. If you make a <u>solution</u> of calcium hydroxide in water (called <u>limewater</u>) and bubble gas through it, the solution will turn <u>cloudy</u> if there's <u>carbon dioxide</u> in the gas. The cloudiness is caused by the formation of <u>calcium carbonate</u>.

> **calcium hydroxide + carbon dioxide → calcium carbonate + water**
> $$Ca(OH)_2 + CO_2 \rightarrow CaCO_3 + H_2O$$

Using Limestone

Limestone is really very handy. However, digging huge amounts of limestone out of the ground can have a quite a significant negative effect on the environment.

Limestone is Used to Make *Other Useful Things* Too

1) Powdered limestone is <u>heated</u> in a kiln with <u>powdered clay</u> to make <u>cement</u>.

2) Cement can be mixed with <u>sand</u> and <u>water</u> to make <u>mortar</u>. <u>Mortar</u> is the stuff you stick <u>bricks</u> together with. You can also add <u>calcium hydroxide</u> to mortar.

3) Or you can mix cement with <u>sand</u> and <u>aggregate</u> (water and gravel) to make <u>concrete</u>.

Quarrying Limestone Makes a *Right Mess* of the *Landscape*

Digging limestone out of the ground can cause environmental problems.

1) For a start, it makes <u>huge ugly holes</u> which permanently change the landscape.

2) <u>Quarrying</u> processes, like blasting rocks apart with explosives, make lots of <u>noise</u> and <u>dust</u> in quiet, scenic areas.

3) Quarrying <u>destroys the habitats</u> of animals and birds.

4) The limestone needs to be <u>transported away</u> from the quarry — usually in lorries. This causes more noise and pollution.

5) Waste materials produce unsightly <u>tips</u>.

Limestone's amazingly useful

Wow. It sounds like you can achieve <u>pretty much anything</u> with limestone, possibly apart from a bouncy castle. I wonder what we'd be using instead if all those sea creatures hadn't died and conveniently become rock? But don't forget that quarrying is a messy business.

Using Limestone

So using limestone ain't all hunky-dory — making stuff from it causes quite a few problems.

Making Stuff From Limestone Causes Pollution Too

1) <u>Cement factories</u> make a lot of <u>dust</u>, which can cause <u>breathing problems</u> for some people.

2) <u>Energy</u> is needed to produce cement and quicklime.
The energy is likely to come from burning <u>fossil fuels</u>, which causes pollution.

See pages 91-92 for more on pollution caused by burning fossil fuels.

But on the Plus Side...

1) Limestone provides things that people want — like <u>houses</u> and <u>roads</u>. Chemicals used in making <u>dyes</u>, <u>paints</u> and <u>medicines</u> also come from limestone.

2) Limestone products are used to <u>neutralise acidic soil</u>. Acidity in lakes and rivers caused by <u>acid rain</u> is also <u>neutralised</u> by limestone products.

3) Limestone is also used in power station chimneys to <u>neutralise sulfur dioxide</u>, which is a cause of acid rain.

4) The quarry and associated businesses provide <u>jobs</u> for people and bring more money into the <u>local economy</u>. This can lead to <u>local improvements</u> in transport, roads, recreation facilities and health.

5) Once quarrying is complete, <u>landscaping</u> and <u>restoration</u> of the area is normally required as part of the planning permission.

Limestone Products Have Advantages and Disadvantages

1) Limestone and concrete (made from cement) are used as <u>building materials</u>. In some cases they're <u>perfect</u> for the job, but in other cases they're a bit of a compromise.

2) Limestone is <u>widely available</u> and is <u>cheaper</u> than granite or marble. It's also a fairly easy rock to <u>cut</u>.

3) Some limestone is more <u>hard-wearing</u> than marble, but it still looks <u>attractive</u>.

4) Concrete can be poured into <u>moulds</u> to make blocks or panels that can be joined together. It's a <u>very quick and cheap</u> way of constructing buildings — <u>and it shows</u>... — concrete has got to be the most <u>hideously unattractive</u> building material ever known.

5) Limestone, concrete and cement <u>don't rot</u> when they get wet like wood does. They can't be gnawed away by <u>insects</u> or <u>rodents</u> either. And to top it off, they're <u>fire-resistant</u> too.

6) Concrete <u>doesn't corrode</u> like lots of metals do. It does have a fairly <u>low tensile strength</u> though, and can crack. If it's <u>reinforced</u> with steel bars it'll be much stronger.

Tough revision here — this stuff's rock hard...

There's a <u>downside</u> to everything, including using limestone — ripping open huge quarries definitely <u>spoils the countryside</u>. But you have to find a <u>balance</u> between the environmental and ecological factors and the economic and social factors — is it worth keeping the countryside pristine if it means loads of people have nowhere to live because there's no stuff available to build houses with?

Salt

In <u>hot countries</u> they get salt by pouring <u>sea water</u> into big flat open tanks and letting the <u>Sun</u> evaporate the water, leaving the salt behind. This is no good in Britain though — there isn't enough sunshine.

Salt is Left Behind by Evaporation

1) In <u>Britain</u>, salt is extracted from <u>underground deposits</u>.

2) These underground deposits were formed when <u>ancient seas</u> containing dissolved salt <u>evaporated</u>.

3) The salt that was left behind was buried and compressed by other layers of sediment over millions of years.

4) There are massive deposits of this <u>rock salt</u> under <u>Cheshire</u> and <u>Teeside</u>.

Salt Can Also be Obtained from the Sea

1) In hot countries like Australia and China salt can be obtained by <u>evaporating seawater</u>.

2) Seawater flows into specially built <u>shallow pools</u>.

3) It is left to evaporate in the <u>sun</u>, leaving the salt behind.

4) This process is <u>repeated</u> several times and then the salt is <u>collected</u>.

5) This method produces the <u>purest salt</u> — it can be nearly 100% sodium chloride.

Salt

Rock Salt is Extracted by Mining

1) Rock salt is a mixture of salt and impurities.
2) It can be extracted by normal mining or by solution mining.

Normal Salt Mining involves Physical Extraction of Rock Salt

1) Rock salt is drilled, blasted and dug out and brought to the surface using machinery.

2) Most rock salt obtained through this type of mining is used on roads to stop ice forming.

> - The salt in the mixture melts ice by lowering the freezing point of water to around –5 °C.
> - The sand and grit give grip on unmelted ice.

3) The salt can also be separated out and used to enhance the flavour in food or for making chemicals.

Solution Mining is an Alternative Method of Salt Mining

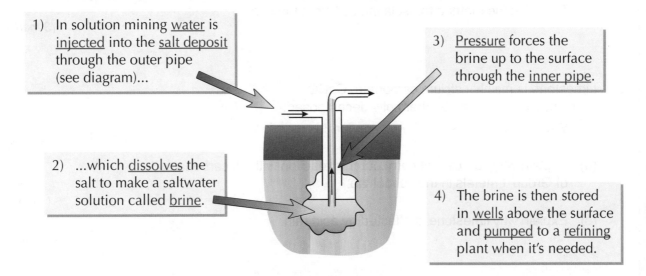

1) In solution mining water is injected into the salt deposit through the outer pipe (see diagram)...

2) ...which dissolves the salt to make a saltwater solution called brine.

3) Pressure forces the brine up to the surface through the inner pipe.

4) The brine is then stored in wells above the surface and pumped to a refining plant when it's needed.

5) Impurities are removed from the brine in the refining plant and it's then pumped into containers. The brine is then boiled to make the water evaporate, leaving the salt behind.

6) Most table salt and a lot of the salt used for chemical production is produced this way.

Just reading this page is making me thirsty...

Plenty of ways to get your hands on some salt then. Remember, salt isn't just used to flavour food — rock salt helps keep the roads safe in winter and salt's used in all sorts of chemical processes.

Warm-Up and Exam Questions

Ah, that time again... time to show off your accumulated knowledge by answering some questions.

Warm-Up Questions

1) Give two examples of materials that occur naturally.
2) Name one synthetic material.
3) Give two things that limestone is used for.
4) Give two advantages of limestone building materials.
5) Salt can be obtained by mining or evaporation.
   Which method produces the purest salt?

Exam Questions

1   Salt can be obtained as sea salt or as rock salt.
     (a) (i)   Explain what rock salt is.

(1 mark)

          (ii) Describe how solution mining can be used to obtain rock salt.

(4 marks)

     (b) Describe the method for isolating salt from the sea.

(3 marks)

2   Limestone is mainly calcium carbonate, $CaCO_3$.
     Calcium carbonate can be thermally decomposed.
     (a) Write a balanced symbol equation for this reaction.

(1 mark)

     (b) Explain why you cannot carry out this reaction with all carbonates
          of Group 1 metals in the school lab.

(1 mark)

     (c) Explain why limestone is affected by acid rain.

(2 marks)

3   Limestone is often used to make building materials.
     (a) How is cement made from limestone?

(2 marks)

     (b) How is concrete made from limestone?

(2 marks)

     (c) *In this question you will be assessed on the quality of your English, the organisation
          of your ideas and your use of appropriate specialist vocabulary.*
          Outline the negative impacts of quarrying limestone and using
          it to produce building materials.

(6 marks)

Getting Metals From Rocks

A few <u>unreactive metals</u> like <u>gold</u> are found in the Earth as the <u>metal itself</u>, rather than as a compound. The rest of the metals we get by extracting them from rocks — and I bet you're just itching to find out how...

Ores Contain **Enough Metal** to Make **Extraction** *Worthwhile*

1) A <u>metal ore</u> is a <u>rock</u> which contains <u>enough metal</u> to make it <u>worthwhile</u> extracting the metal from it.

2) In many cases the ore is an <u>oxide</u> of the metal. For example, the main <u>aluminium ore</u> is called <u>bauxite</u> — it's aluminium oxide (Al_2O_3).

3) <u>Most metals</u> need to be extracted from their ores using a <u>chemical reaction</u>.

4) The <u>economics</u> (profitability) of metal extraction can <u>change</u> over <u>time</u>. For example:

> * If the market <u>price</u> of a metal <u>drops</u> a lot, it <u>might not</u> be worth extracting it. If the <u>price increases</u> a lot then it <u>might be worth</u> extracting <u>more</u> of it.
>
> * As <u>technology improves</u>, it becomes possible to <u>extract more</u> metal from a sample of rock than was originally possible. So it might now be <u>worth</u> extracting metal that <u>wasn't</u> worth extracting <u>in the past</u>.

Metals Are **Extracted** *From their Ores* **Chemically**

1) A metal can be extracted from its ore <u>chemically</u> — by <u>reduction</u> (see page 38) or by <u>electrolysis</u> (splitting with electricity, see pages 39-40).

2) Some ores may have to be <u>concentrated</u> before the metal is extracted — this just involves getting rid of the <u>unwanted rocky material</u>.

3) <u>Electrolysis</u> can also be used to <u>purify</u> the extracted metal (see page 39).

Occasionally, some metals are extracted from their ores using displacement reactions (see page 41).

You've got to keep your mind on the money

Extracting metals is all about <u>money</u>. If extracting a metal from a rock isn't going to make any cash then quite frankly not many people are interested. That's just the way it is.

The Reactivity Series

How easy it is to get a metal out of its ore all comes down to the metal's position in the reactivity series.

Some Metals can be Extracted by Reduction with Carbon

1) A metal can be extracted from its ore chemically by reduction using carbon.

2) When an ore is reduced, oxygen is removed from it, e.g.

$$2Fe_2O_3 \quad + \quad 3C \quad \rightarrow \quad 4Fe \quad + \quad 3CO_2$$

iron(III) oxide + carbon → iron + carbon dioxide

3) The position of the metal in the reactivity series determines whether it can be extracted by reduction with carbon.

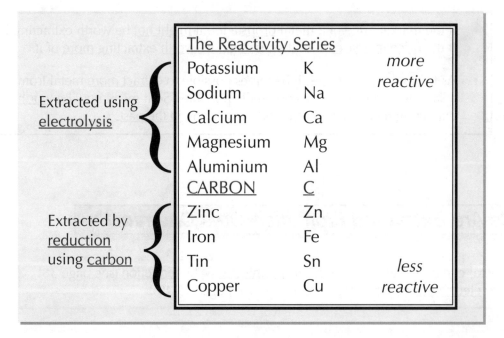

The Reactivity Series

Extracted using electrolysis

Potassium — K — more reactive
Sodium — Na
Calcium — Ca
Magnesium — Mg
Aluminium — Al
CARBON — C

Extracted by reduction using carbon

Zinc — Zn
Iron — Fe
Tin — Sn — less reactive
Copper — Cu

- Metals higher than carbon in the reactivity series have to be extracted using electrolysis, which is expensive.

- Metals below carbon in the reactivity series can be extracted by reduction using carbon. For example, iron oxide is reduced in a blast furnace to make iron.
  This is because carbon can only take the oxygen away from metals which are less reactive than carbon itself is.

Extraction of metals is difficult

Extracting metals isn't cheap. You have to pay for special equipment, energy and labour. Then there's the cost of getting the ore to the extraction plant. If there's a choice of extraction methods, a company always picks the cheapest, unless there's a good reason not to — they're not extracting it for fun.

Extraction of Metals

You may think you know all you could ever want to know about how to get metals from rocks, but no — there's <u>more</u> of it. Think of each of the facts on this page as a little <u>gold nugget</u>. Or, er, a copper one.

Some Metals have to be Extracted by Electrolysis

1) Metals that are <u>more reactive</u> than carbon (see page 38) have to be extracted using electrolysis of <u>molten compounds</u>.

2) An example of a metal that has to be extracted this way is <u>aluminium</u>.

3) However, the process is <u>much more expensive</u> than reduction with carbon (see page 38) because it <u>uses a lot of energy</u>.

> <u>FOR EXAMPLE</u>: a <u>high temperature</u> is needed to <u>melt</u> aluminium oxide so that <u>aluminium</u> can be extracted — this requires a lot of <u>energy</u>, which makes it an <u>expensive</u> process.

Copper is Purified by Electrolysis

A copper ore

1) Copper can be easily extracted by <u>reduction with carbon</u> (see page 38). The ore is <u>heated</u> in a <u>furnace</u> — this is called <u>smelting</u>.

2) However, the copper produced this way is <u>impure</u> — and impure copper <u>doesn't</u> conduct electricity very well. This <u>isn't</u> very <u>useful</u> because a lot of copper is used to make <u>electrical wiring</u>.

3) So <u>electrolysis</u> is also used to <u>purify</u> it, even though it's quite <u>expensive</u>.

4) This produces <u>very pure</u> copper, which is a <u>much better conductor</u>.

You could <u>extract</u> copper straight from its ore by electrolysis if you wanted to, but it's more expensive than using reduction with carbon.

Electrolysis is expensive...

...but it's also pretty <u>useful</u> when it comes to getting hold of metals. It's used to <u>extract</u> some metals and to <u>purify</u> others — now that's what I call multitasking. Just think where we'd be without quality <u>copper wire</u> to conduct <u>electricity</u> — in the dark, for a start.

Electrolysis

Here's a bit of the detail about how <u>electrolysis</u> is used to purify metals.

Electrolysis Means "Splitting Up with Electricity"

1) <u>Electrolysis</u> is the <u>breaking down</u> of a substance using <u>electricity</u>.

2) It requires a <u>liquid</u> to <u>conduct</u> the <u>electricity</u>, called the <u>electrolyte</u>.

3) Electrolytes are often <u>metal salt solutions</u> made from the ore
(e.g. copper sulfate) or <u>molten metal oxides</u>.

4) The electrolyte has <u>free ions</u> — these <u>conduct</u> the electricity and allow the whole thing to work.

5) Electrons are <u>taken away</u> by the <u>positive electrode (anode)</u> and <u>given away</u> by the <u>negative electrode</u>
<u>(cathode)</u>. As ions gain or lose electrons they become atoms or molecules and are released.

Here's how electrolysis is used to get <u>copper</u>:

1) <u>Electrons</u> are <u>pulled off</u> copper atoms at the <u>anode</u>,
causing them to go into solution as <u>Cu^{2+} ions</u>.

2) <u>Cu^{2+} ions</u> near the <u>cathode</u> gain electrons and turn back into <u>copper atoms</u>.

3) The <u>impurities</u> are dropped at the <u>anode</u> as a <u>sludge</u>,
whilst <u>pure copper atoms</u> bond to the <u>cathode</u>.

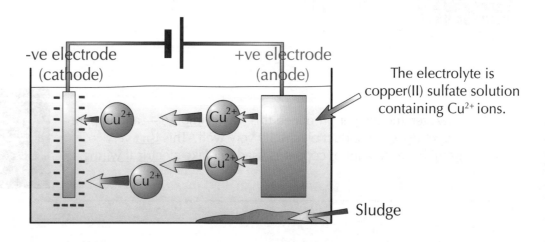

-ve electrode
(cathode)

+ve electrode
(anode)

The electrolyte is
copper(II) sulfate solution
containing Cu^{2+} ions.

Cu^{2+}

Cu^{2+}

Cu^{2+}

Cu^{2+}

Sludge

The <u>cathode</u> starts as
a <u>thin</u> piece of <u>pure</u>
<u>copper</u> and more pure
copper <u>adds</u> to it.

The <u>anode</u> is just a big
lump of <u>impure copper</u>,
which will <u>dissolve</u>.

Copper is a really useful metal

The skin of the <u>Statue of Liberty</u> is made of copper — about 80 tonnes of it in fact.
Its surface reacts with gases in the air to form <u>copper carbonate</u> — which is why it's
that pretty shade of <u>green</u>. It was a present from France to the United States —
I wonder if they found any wrapping paper big enough?

Extracting Copper

Don't stop now — there's still a bit more about copper extraction... sigh, it's a hard life.

A *Displacement Reaction* Can Be Used to *Extract Copper*

1) More reactive metals react more vigorously than less reactive metals.

2) If you put a reactive metal into a solution of a dissolved metal compound, the reactive metal will replace the less reactive metal in the compound.

3) This is because the more reactive metal bonds more strongly to the non-metal bit of the compound and pushes out the less reactive metal.

4) For example, scrap iron can be used to displace copper from solution — this is really useful because iron is cheap but copper is expensive. If some iron is put in a solution of copper sulfate, the more reactive iron will "kick out" the less reactive copper from the solution. You end up with iron sulfate solution and copper metal.

$$\text{copper sulfate} + \text{iron} \rightarrow \text{iron sulfate} + \text{copper}$$

5) If a piece of silver metal is put into a solution of copper sulfate, nothing happens. The more reactive metal (copper) is already in the solution.

Copper-rich Ores are in Short Supply

1) The supply of copper-rich ores is limited, so it's important to recycle as much copper as possible.

2) The demand for copper is growing and this may lead to shortages in the future.

3) Scientists are looking into new ways of extracting copper from low-grade ores (ores that only contain small amounts of copper) or from the waste that is currently produced when copper is extracted.

4) Examples of new methods to extract copper are bioleaching and phytomining:

Bioleaching

This uses bacteria to separate copper from copper sulfide. The bacteria get energy from the bond between copper and sulfur, separating out the copper from the ore in the process. The leachate (the solution produced by the process) contains copper, which can be extracted, e.g. by filtering.

Phytomining

This involves growing plants in soil that contains copper. The plants can't use or get rid of the copper so it gradually builds up in the leaves. The plants can be harvested, dried and burned in a furnace. The copper can be collected from the ash left in the furnace.

5) Traditional methods of copper mining are pretty damaging to the environment (see page 42). These new methods of extraction have a much smaller impact, but the disadvantage is that they're slow.

Personally, I'd rather be pound rich than copper rich...

The fact that copper-rich ore supplies are dwindling means that scientists have to come up with ever-more-cunning methods to extract it. It also gives you something else to learn about.

Impacts of Extracting Metals

Metals are very useful. Just imagine if all knives and forks were made of plastic instead — there'd be prongs snapping all over the place at dinner time. However, mining for metals isn't all good news...

Mining *can be* Good...

1) Ores are finite resources. This means that there's a <u>limited amount</u> of them — eventually, they'll run out.

2) People have to balance the <u>social</u>, <u>economic</u> and <u>environmental</u> effects of mining the ores.

3) So, mining metal ores is <u>good</u> because <u>useful products</u> can be made. It also provides local people with <u>jobs</u> and brings <u>money</u> into the area. This means services such as <u>transport</u> and <u>health</u> can be improved.

...*or* Bad

1) But mining ores is <u>bad for the environment</u> as it uses loads of energy, scars the landscape and destroys habitats. Also, noise, dust and pollution are caused by an increase in traffic.

2) Deep mine shafts can also be <u>dangerous</u> for a long time after the mine has been abandoned.

3) Land above disused mines can <u>collapse into the holes</u> — this is called <u>subsidence</u>. Subsidence can affect buildings near the mines, including people's homes.

4) The risk of subsidence is <u>reduced</u> by leaving <u>well-supported</u> caverns in mines (e.g. supported by pillars of rock). Also, caverns can be spaced well apart and <u>filled in</u> when no longer used.

Impacts of Extracting Metals

Once metals are finished with, it's better to recycle them than to dig up more ore and extract new metal.

Recycling metals is important

1) Mining and extracting metals takes lots of <u>energy</u>, most of which comes from burning <u>fossil fuels</u>.

2) Fossil fuels are <u>running out</u> so it's important to <u>conserve</u> them. Not only this, but burning them contributes to <u>acid rain</u>, <u>global dimming</u> and <u>climate change</u> (see pages 91 and 93).

3) Recycling metals only uses a <u>small fraction</u> of the energy needed to mine and extract new metal. E.g. recycling copper only takes 15% of the energy that's needed to mine and extract new copper.

4) Energy doesn't come cheap, so recycling <u>saves money</u> too.

5) Also, there's a <u>finite amount</u> of each <u>metal</u> in the Earth. Recycling conserves these resources.

6) Recycling metal cuts down on the amount of rubbish that gets sent to <u>landfill</u>. Landfill takes up space and <u>pollutes</u> the surroundings. If all the aluminium cans in the UK were recycled, there'd be 14 million fewer dustbins to empty each year.

But...

...if you didn't recycle, say, <u>aluminium</u>, you'd have to <u>mine</u> more aluminium ore — <u>4 tonnes</u> for every <u>1 tonne</u> of aluminium you need. But mining makes a mess of the <u>landscape</u> (and these mines are often in <u>rainforests</u>). The ore then needs to be <u>transported</u>, and the aluminium <u>extracted</u> (which uses <u>loads</u> of electricity). And don't forget the cost of sending your <u>used</u> aluminium to <u>landfill</u>.

So it's a <u>complex</u> calculation, but for every 1 kg of aluminium cans you recycle, you <u>save</u>:

- <u>95%</u> or so of the <u>energy</u> needed to mine and extract 'fresh' aluminium,
- <u>4 kg</u> of aluminium ore,
- a <u>lot</u> of waste.

In fact, aluminium's about the most cost-effective metal to recycle.

Get back on your bike again — recycle...

Recycling metals saves <u>natural resources</u> and <u>money</u> and reduces <u>environmental problems</u>. It's great. There's no limit to the number of times metals like aluminium, copper and steel can be recycled. So your humble little drink can may one day form part of a powerful robot who takes over the galaxy.

Warm-Up and Exam Questions

You've arrived at the next set of warm-up and exam questions. It's really important to find out what you know (as well as what you think you know but actually don't). So give them a go.

Warm-Up Questions

1) What is an ore?
2) Name a metal which can be extracted from its ore by reduction with carbon.
3) Why is electrolysis expensive?
4) Explain how phytomining can be used to extract copper.
5) Give one reason why mining ores is bad for the environment.

Exam Questions

1   Copper is not usually extracted from its ore by electrolysis.
    (a) Suggest why this is.

(2 marks)

    (b) Copper sulfate solution can be electrolysed to obtain pure copper.
         Explain what happens to the Cu^{2+} ions in the solution during this process.

(2 marks)

2   Copper needs to be extracted from its ore before it can be used.
    (a) Why are scientists trying to find new ways to extract copper
         from low-grade ores?

(1 mark)

    (b) It is possible to extract copper from copper sulfide using bacteria.
         (i)   What is the name of this method?

(1 mark)

         (ii) Describe the process involved in this method.

(1 mark)

         (iii) Give **one** advantage of using this method rather than other methods.

(1 mark)

         (iv) Give **one** disadvantage of using this method rather than other methods.

(1 mark)

3   *In this question you will be assessed on the quality of your English, the organisation of your ideas and your use of appropriate specialist vocabulary.*

    Mining ores has social, economic and environmental effects.
    Discuss the positive and negative effects of mining metal ores.

(6 marks)

Exam Questions

4   The diagram shows part of the reactivity series of metals, together with carbon.

Potassium	K	*more*
Sodium	Na	*reactive*
Calcium	Ca	
Magnesium	Mg	
Aluminium	Al	
CARBON	C	
Zinc	Zn	
Iron	Fe	
Tin	Sn	*less*
Copper	Cu	*reactive*

(a)   Name one metal which is extracted from its ore using electrolysis.

(1 mark)

(b)   Some metals can be extracted from their ores by reduction with carbon, producing the metal and carbon dioxide.

   (i)   Explain the meaning of reduction.

(1 mark)

   (ii)   Write a word equation for the reduction of zinc oxide by carbon.

(1 mark)

(c)   Iron can be extracted by the reduction of iron(III) oxide (Fe_2O_3) with carbon (C), to produce iron and carbon dioxide.
      Write a balanced symbol equation for this reaction.

(2 marks)

(d)   (i)   In which of these test tubes will a reaction occur?

(1 mark)

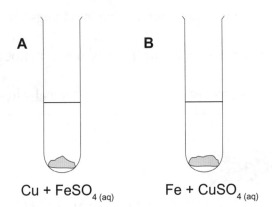

A   B

Cu + FeSO$_{4 (aq)}$ Fe + CuSO$_{4 (aq)}$

   (ii)   Explain your answer.

(2 marks)

(e)   Explain why recycling metals is important.

(6 marks)

Properties of Metals

Metals are all the same but slightly different. They have some basic properties in common, but each has its own specific combination of properties, which mean you use different ones for different purposes.

Metals are Strong and Bendy and They're Great Conductors

1) Most of the elements are metals — so they cover most of the periodic table. In fact, only the elements on the far right are non-metals.

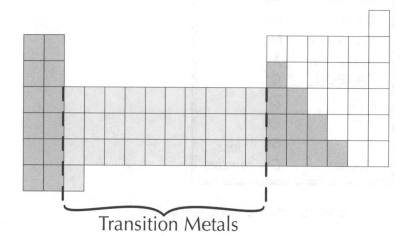

Transition Metals

The coloured elements are metals. Just look at them all — there's loads of them!

2) All metals have some fairly similar basic properties:

> - Metals are strong (hard to break), but they can be bent or hammered into different shapes.
> - They're great at conducting heat.
> - They conduct electricity well.

3) Metals (and especially transition metals, which are found in the centre block of the periodic table) have loads of everyday uses because of these properties...

> - Their strength and 'bendability' makes them handy for making into things like bridges and car bodies.
> - Metals are ideal if you want to make something that heat needs to travel through, like a saucepan base.
> - And their conductivity makes them great for making things like electrical wires.

Metals are Good — but Not Perfect

1) Metals are very useful structural materials, but some corrode when exposed to air and water, so they need to be protected, e.g. by painting. If metals corrode, they lose their strength and hardness.

2) Metals can get 'tired' when stresses and strains are repeatedly put on them over time. This is known as metal fatigue and leads to metals breaking, which can be very dangerous, e.g. in planes.

Metals have lots of properties in common

So, all metals conduct electricity and heat and can be bent into shape. This makes them pretty useful. Don't go thinking metals are perfect though — they're not immune to corrosion and fatigue.

Properties of Metals

Some metals like titanium, aluminium and copper have special properties.

A Metal's **Exact Properties** Decide How It's Best **Used**

1) The properties on the previous page are <u>typical properties</u> of metals.
   Not all metals are the same though...

> <u>Copper</u> is a <u>good conductor</u> of <u>electricity</u>, so it's ideal for drawing out into electrical wires.
> It's <u>hard</u> and <u>strong</u> but can be <u>bent</u>. It also <u>doesn't react with water</u>.

> <u>Aluminium</u> is <u>corrosion-resistant</u> and has a <u>low density</u>.
> Pure aluminium <u>isn't</u> particularly strong, but it forms
> hard, strong alloys (see page 49).

> <u>Titanium</u> is another <u>low density metal</u>.
> Unlike aluminium it's <u>very strong</u>.
> It is also <u>corrosion-resistant</u>.

2) <u>Different metals</u> are chosen for <u>different uses</u> because of their specific properties.
   For example:

- If you were doing some <u>plumbing</u>, you'd pick a
  metal that could be <u>bent</u> to make pipes and tanks,
  and is below hydrogen in the reactivity series so it
  <u>doesn't react with water</u>. <u>Copper</u> is great for this.

- If you wanted to make an <u>aeroplane</u>, you'd probably use
  metal as it's <u>strong</u> and can be <u>bent into shape</u>. But you'd also
  need it to be <u>light</u>, so <u>aluminium</u> would be a good choice.

- And if you were making <u>replacement hips</u>, you'd pick a
  metal that <u>won't corrode</u> when it comes in contact with
  water. It'd also have to be <u>light</u> too, and not too bendy.
  <u>Titanium</u> has all of these properties so it's used for this.

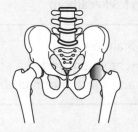

Aluminium isn't just used for drinks cans

The general idea here is that metals have different <u>properties</u>, so they're suitable for different <u>tasks</u>.
If you're choosing which metal to make something from, it's all about finding the right one for the job...

Alloys

Pure metals often aren't quite right for certain jobs. So scientists mix two metals together (or mix a metal with a non-metal) — creating an alloy with the properties they want.

Pure Iron Tends to be a Bit Too Bendy

1) 'Iron' straight from the blast furnace (see page 38) is only 96% iron. The other 4% is impurities such as carbon.

2) This impure iron is used as cast iron. It's handy for making ornamental railings, but it doesn't have many other uses because it's brittle.

3) So all the impurities are removed from most of the blast furnace iron. This pure iron has a regular arrangement of identical atoms. The layers of atoms can slide over each other, which makes the iron soft and easily shaped. This iron is far too bendy for most uses.

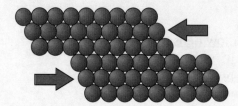

Most Iron is Converted into Steel — an Alloy

Most of the pure iron is changed into alloys called steels. Steels are formed by adding small amounts of carbon and sometimes other metals to the iron.

TYPE OF STEEL	PROPERTIES	USES
Low carbon steel (0.1% carbon)	easily shaped	car bodies (see page 50)
High carbon steel (1.5% carbon)	very hard, inflexible	blades for cutting tools, bridges
Stainless steel (chromium added, and sometimes nickel)	corrosion-resistant	cutlery, containers for corrosive substances

Most iron is changed into steel, otherwise it's too bendy or too brittle

The Eiffel Tower is made of iron — but the problem with iron is, it goes rusty if air and water get to it. So the Eiffel Tower has to be painted every seven years to make sure that it doesn't rust. This is quite a job and takes an entire year for a team of 25 painters. Too bad they didn't use stainless steel.

Alloys

Alloys are really useful. Lots of the metals we use are alloys...

Alloys are Harder Than Pure Metals

1) Different elements have different sized atoms. So when an element such as carbon is added to pure iron, the smaller carbon atom will upset the layers of pure iron atoms, making it more difficult for them to slide over each other. So alloys are harder.

2) Many metals in use today are actually alloys. E.g.:

BRONZE = COPPER + TIN

Bronze is harder than copper.
It's good for making medals and statues from.

CUPRONICKEL = COPPER + NICKEL

This is hard and corrosion resistant.
It's used to make "silver" coins.

GOLD ALLOYS ARE USED TO MAKE JEWELLERY

Pure gold is too soft. Metals such as zinc, copper, silver, palladium and nickel are used to harden the "gold".

ALUMINIUM ALLOYS ARE USED TO MAKE AIRCRAFT

Aluminium has a low density, but it's alloyed with small amounts of other metals to make it stronger.

3) In the past, the development of alloys was by trial and error. But nowadays we understand much more about the properties of metals, so alloys can be designed for specific uses.

Alloys are really important in industry

If the properties of a metal aren't quite suited to a job, an alloy is often used instead. To make an alloy you mix one metal with another metal or non-metal. The finished alloy can be a lot harder, or less brittle — the properties can be varied and they can be made to suit a particular job really well.

Building Cars

There are loads of different materials in your average car — different materials have different properties and so have different uses. Makes sense.

Iron and Steel **Corrode Much More** than Aluminium

1) Iron corrodes easily. In other words, it rusts.

2) Rusting only happens when the iron's in contact with both oxygen (from the air) and water.

3) The chemical reaction that takes place when iron corrodes is an oxidation reaction. The iron gains oxygen to form iron(III) oxide. Water then becomes loosely bonded to the iron(III) oxide and the result is hydrated iron(III) oxide — which we call rust.

   Here's the word equation for the reaction:

   The word "rust" is only used for the corrosion of iron, not other metals.

iron + oxygen + water → hydrated iron(III) oxide

4) Unfortunately, rust is a soft crumbly solid that soon flakes off to leave more iron available to rust again. And if the water's salty or acidic, rusting will take place a lot quicker.

5) Cars in coastal places rust a lot because they get covered in salty sea-spray. Cars in dry deserty places hardly rust at all.

> Aluminium doesn't corrode when it's wet. This is a bit odd because aluminium is more reactive than iron. What happens is that the aluminium reacts very quickly with oxygen in the air to form aluminium oxide. A nice protective layer of aluminium oxide sticks firmly to the aluminium below and stops any further reaction taking place (the oxide isn't crumbly and flaky like rust, so it won't fall off).

Car Bodies: Aluminium or Steel?

Aluminium has two big advantages over steel:

> 1) It has a much lower density, so the car body of an aluminium car will be lighter than the same car made of steel. This gives the aluminium car much better fuel economy, which saves fuel resources.

> 2) A car body made with aluminium corrodes less and so it'll have a longer lifetime.

But aluminium has a massive disadvantage. It costs a lot more than iron or steel. That's why car manufacturers tend to build cars out of steel instead.

Car Materials and Recycling

Just like cans, glass and paper, cars can also be recycled.

You Need *Various Materials* to Build *Different Bits* of a Car

Steel is strong and it can be hammered into sheets and welded together — good for the bodywork.

Aluminium is strong and has a low density — it's used for parts of the engine, to reduce weight.

Glass is transparent — cars need windscreens and windows.

Plastics are light and hardwearing, so they're used as internal coverings for doors, dashboards, etc. They're also electrical insulators, used for covering electrical wires

Fibres (natural and synthetic) are hard-wearing, so they're used to cover the seats and floor.

Recycling Cars is Important

1) As with all recycling, the idea is to save natural resources, save money and reduce landfill use.

2) At the moment a lot of the metal from a scrap car is recycled, though most of the other materials (e.g. plastics, rubber, etc.) go into landfill.

3) But European laws are now in place saying that 85% of the materials in a car (rising to 95% of a car by 2015) must be recyclable.

4) The biggest problem with recycling all the non-metal bits of a car is that they have to be separated before they can be recycled. Sorting out different types of plastic is a pain in the neck.

CGP jokes — 85% recycled since 1996...

When manufacturers choose materials for cars, they have to weigh up alternatives — they balance safety, environmental impact, and cost. You never know, one day you could be asked to do the same.

Warm-Up and Exam Questions

The warm-up questions run quickly over the basic facts from the last few pages.
Unless you've learnt the facts first you'll find the exam questions pretty difficult.

Warm-Up Questions

1) Give three useful physical properties of most metals.
2) Where are the transition metals found in the periodic table?
3) What is an alloy?
4) Which are harder, pure metals or alloys?
5) Write down the word equation for the reaction in which rust is formed from iron.

Exam Questions

1   The tensile strength of a material can be measured as the force it will withstand before
it breaks. Look at the table below. It shows the cost of some metals as well as their
tensile strength.

Metal	Cost / tonne (£)	Tensile strength force (MPa)
aluminium	1150	45
copper	4600	220
cast iron	180	200
steel	200	400
brass	2700	550
tungsten	24 000	1510

(a) Cast iron has few uses and most of it is turned into steel.

   (i)   What is cast iron?

(1 mark)

   (ii)   Use the table to explain why steel is more useful than cast iron.

(1 mark)

(b) Steel is used to make cables used in the construction of bridges.
Brass is not used to make cables as it is too expensive.

Using data from the table, explain why steel is the most suitable material
to make cables.

(2 marks)

(c) Aluminium is used to make aircraft.
Aircraft need to be strong but aluminium is not a strong metal.

   (i)   Explain how aluminium can be made stronger.

(1 mark)

   (ii)   Using your own knowledge, give two properties of aluminium that make it
suitable for making aircraft.

(2 marks)

Exam Questions

2   Titanium is a transition metal that is used in hip replacements.
    (a) Give three properties of titanium that make it suitable for this use.
                                                                                        (3 marks)
    (b) Give one other property of titanium.
                                                                                        (1 mark)

3   Low carbon steel and high carbon steel are two different alloys of iron.
    (a) Give one difference in the properties of these two alloys.
                                                                                        (2 marks)
    (b) Give one use for each of these alloys.
                                                                                        (2 marks)
    (c) Steel **X** is an alloy of iron that is resistant to corrosion.
         (i)   Name this alloy.
                                                                                        (1 mark)
         (ii) Give one use of steel **X**.
                                                                                        (1 mark)
    (d) Suggest why scientists are now able to design alloys for specific uses.
                                                                                        (1 mark)

4   Cars are manufactured from many different materials.
    (a) Name one material that is commonly used for the bodywork of a car.
                                                                                        (1 mark)
    (b) (i)   Give two reasons why aluminium is used rather than steel to
               make car engines.
                                                                                        (2 marks)
         (ii) Give one disadvantage of using aluminium rather than steel to
               make car engines.
                                                                                        (1 mark)
    (c) Explain why it is important that to ensure that car parts can be recycled.
                                                                                        (3 marks)

5   Alloys are often used instead of pure metals because
    A   they are more plentiful.
    B   their properties make them more suitable for the application.
    C   their melting points are higher.
    D   they are completely inert.
                                                                                        (1 mark)

Revision Summary for Section 2

Okay, if you were just about to turn the page without doing these revision summary questions, then stop. What kind of attitude is that... Is that really the way you want to live your life... running, playing and having fun... Of course not. That's right. Do the questions. It's for the best all round.

1) Name a material we use that comes from:
   a) plants   b) animals

2) What is the advantage of synthetic rubber over natural rubber?

3) What products are produced when limestone reacts with an acid?

4) What is calcium hydroxide used for?

5) Name three building materials made from limestone.

6) Plans to develop a limestone quarry and a cement factory on some hills next to your town are announced. Describe the views that the following might have:
   a) dog owners
   b) a mother of young children
   c) the owner of a cafe
   d) a beetle

7) Give three methods of obtaining salt industrially.

8) Explain why zinc can be extracted by reduction with carbon but magnesium can't.

9) What is electrolysis?

10) Describe the process of purifying copper by electrolysis.

11) Describe how scrap iron is used to displace copper from solution.

12) What is the name of the method where plants are used to extract metals from soil?

13) Give four advantages of recycling metals rather than digging up new metal ore.

14) Briefly describe two problems with metals.

15) What is the problem with using
   a) iron straight from the blast furnace,
   b) very pure iron?

16) Give two examples of alloys and say what's in them.

17) Write down the word equation for the corrosion of iron.

18) Explain why a car parked on the Brighton seafront rusts more than a car parked in hot, dry Cairo.

19) Why doesn't aluminium corrode when it's wet?

Fractional Distillation of Crude Oil

Crude oil is formed from the buried remains of plants and animals — it's a fossil fuel. Over millions of years, the remains turn to crude oil, which can be extracted by drilling and pumping.

Crude Oil is a *Mixture* of *Hydrocarbons*

1) A mixture consists of two (or more) elements or compounds that aren't chemically bonded to each other.

2) Crude oil is a mixture of many different compounds. Most of the compounds are hydrocarbon molecules.

3) Hydrocarbons are basically fuels such as petrol and diesel. They're made of just carbon and hydrogen.

4) There are no chemical bonds between the different parts of a mixture, so the different hydrocarbon molecules in crude oil aren't chemically bonded to one another.

5) This means that they all keep their original properties, such as their condensing points. The properties of a mixture are just a mixture of the properties of the separate parts.

6) The parts of a mixture can be separated out by physical methods, e.g. crude oil can be split up into its separate fractions by fractional distillation. Each fraction contains molecules with a similar number of carbon atoms to each other (see page 56).

Crude Oil is *Split* into *Separate Groups of Hydrocarbons*

The fractionating column works continuously, with heated crude oil piped in at the bottom. The vaporised oil rises up the column and the various fractions are constantly tapped off at the different levels where they condense.

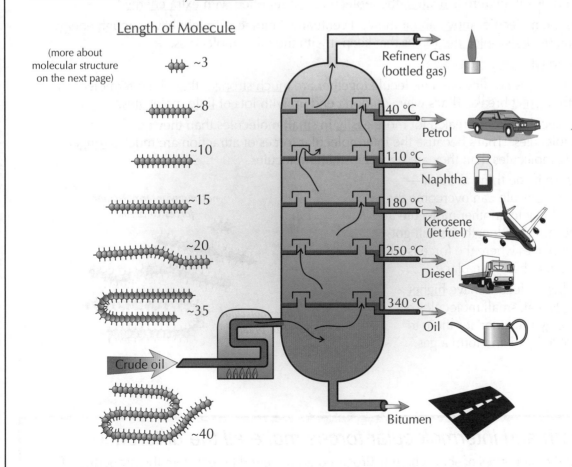

Length of Molecule

(more about molecular structure on the next page)

~3

~8

~10

~15

~20

~35

Crude oil

~40

Refinery Gas (bottled gas)

40 °C — Petrol

110 °C — Naphtha

180 °C — Kerosene (Jet fuel)

250 °C — Diesel

340 °C — Oil

Bitumen

Properties of Crude Oil

The different fractions of crude oil have different properties, and it's all down to their structure.

Hydrocarbon *Properties Change* as the Chain Gets *Longer*

As the length of the carbon chain changes, the properties of the hydrocarbon change.

1) The shorter the molecules, the more runny the hydrocarbon is — that is, the less viscous (gloopy) it is.

2) The shorter the molecules, the more volatile they are. "More volatile" means they turn into a gas at a lower temperature. So, the shorter the molecules, the lower the temperature at which that fraction vaporises or condenses — and the lower its boiling point.

3) Also, the shorter the molecules, the more flammable (easier to ignite) the hydrocarbon is.

The *Properties* Depend on the *Forces Between Molecules*

It's all down to the forces in between hydrocarbons...

1) There are two important types of bond in crude oil:

> a) The strong covalent bonds between the carbons and hydrogens within each hydrocarbon molecule.
> b) The intermolecular forces of attraction between different hydrocarbon molecules in the mixture.

2) When the crude oil mixture is heated, the molecules are supplied with extra energy.

3) This makes the molecules move about more. Eventually a molecule might have enough energy to overcome the intermolecular forces that keep it with the other molecules.

4) It can now go whizzing off as a gas.

5) The covalent bonds holding each molecule together are much stronger than the intermolecular forces, so they don't break. That's why you don't end up with lots of little molecules.

6) The intermolecular forces break a lot more easily in small molecules than they do in bigger molecules. That's because the intermolecular forces of attraction are much stronger between big molecules than they are between small molecules.

7) It makes sense if you think about it — even if a big molecule can overcome the forces attracting it to another molecule at a few points along its length, it's still got lots of other places where the force is still strong enough to hold it in place.

8) That's why big molecules have higher boiling points than small molecules do — more energy is needed for them to break out of a liquid and form a gas.

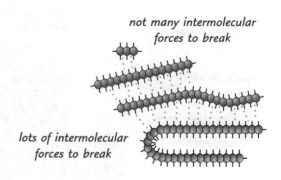

not many intermolecular forces to break

lots of intermolecular forces to break

Chain length and intermolecular forces make all the difference...

Remember that the properties of short-chain hydrocarbons are very different from the properties of long-chain ones — they're less viscous, more volatile and easier to ignite than long-chain hydrocarbons.

Using Crude Oil

Crude oil is a really useful fuel that we use every day.

The **Uses** of Hydrocarbons Depend on Their **Properties**

1) The <u>volatility</u> helps decide what the fraction is used for. The <u>refinery gas fraction</u> has the shortest molecules, so it has the <u>lowest boiling point</u> — in fact it's a gas at room temperature. This makes it ideal for using as <u>bottled gas</u>. It's stored under pressure as liquid in 'bottles'. When the tap on the bottle is opened, the fuel vaporises and flows to the burner where it's ignited.

2) The <u>petrol</u> fraction has longer molecules, so it has a higher boiling point. Petrol is a <u>liquid</u> which is ideal for storing in the fuel tank of a car. It can flow to the engine where it's easily <u>vaporised</u> to mix with the air before it is ignited.

3) The <u>viscosity</u> also helps decide how the hydrocarbons are <u>used</u>. The really gloopy, viscous hydrocarbons are used for <u>lubricating engine parts</u> and for <u>covering roads</u>.

Crude Oil Provides an **Important Fuel** for **Modern Life**

1) Crude oil fractions burn cleanly so they make good <u>fuels</u>. Most modern transport is fuelled by a crude oil fraction, e.g. cars, boats, trains and planes. Parts of crude oil are also burned in <u>central heating systems</u> in homes and in <u>power stations</u> to <u>generate electricity</u>.

2) There's a <u>massive industry</u> with scientists working to find oil reserves, take it out of the ground, and turn it into useful products. As well as fuels, crude oil also provides the raw materials for making various <u>chemicals</u>, including <u>plastics</u>.

3) Often, <u>alternatives</u> to using crude oil fractions as fuel are possible. E.g. electricity can be generated by <u>nuclear</u> power or <u>wind</u> power, there are <u>ethanol</u>-powered cars, and <u>solar</u> energy can be used to heat water.

4) But things tend to be <u>set up</u> for using oil fractions. For example, cars are designed for <u>petrol or diesel</u> and it's <u>readily available</u>. There are filling stations all over the country, with storage facilities and pumps specifically designed for these crude oil fractions. So crude oil fractions are often the <u>easiest and cheapest</u> thing to use.

5) Crude oil fractions are often <u>more reliable</u> too — e.g. solar and wind power won't work without the right weather conditions. Nuclear energy is reliable, but there are lots of concerns about its <u>safety</u> and the storage of radioactive waste.

Using Crude Oil

Nothing as amazingly useful as crude oil would be without its problems.
No, that'd be too good to be true.

Crude Oil Might **Run Out** One Day... **Eeek**

1) Most scientists think that oil will <u>run out</u> — it's a <u>non-renewable fuel</u>.

2) No one knows exactly when it'll run out but there have been heaps of <u>different predictions</u> — e.g. about 40 years ago, scientists predicted that it'd all be gone by the year 2000.

3) <u>New oil reserves</u> are discovered from time to time and <u>technology</u> is constantly improving, so it's now possible to extract oil that was once too <u>difficult</u> or <u>expensive</u> to extract.

4) In the <u>worst-case scenario</u>, oil may be pretty much gone in about 25 years — and that's not far off.

5) Some people think we should <u>immediately stop</u> using oil for things like transport, for which there are alternatives, and keep it for things that it's absolutely <u>essential</u> for, like some chemicals and medicines.

6) It will take time to <u>develop</u> alternative fuels that will satisfy all our energy needs (see page 94 for more info). It'll also take time to <u>adapt things</u> so that the fuels can be used on a wide scale. E.g. we might need different kinds of car engines, or special storage tanks built.

7) One alternative is to generate energy from <u>renewable</u> sources — these are sources that <u>won't run out</u>. Examples of renewable energy sources are <u>wind power</u>, <u>solar power</u> and <u>tidal power</u>.

8) So however long oil does last for, it's a good idea to start <u>conserving</u> it and finding <u>alternatives</u> now.

Crude Oil is **Not** the **Environment's** Best Friend

1) <u>Oil spills</u> can happen as the oil is being transported by tanker — this spells <u>disaster</u> for the local environment. <u>Birds</u> get covered in the stuff and are <u>poisoned</u> as they try to clean themselves. Other creatures, like <u>sea otters</u> and <u>whales</u>, are poisoned too.

2) You have to <u>burn oil</u> to release the energy from it. But burning oil is thought to be a major cause of <u>global warming</u>, <u>acid rain</u> and <u>global dimming</u> — see pages 91 and 93.

If oil alternatives aren't developed, we might get caught short...

Crude oil is <u>really important</u> to our lives. Take <u>petrol</u> for instance — at the first whisper of a shortage, there's mayhem. Loads of people dash to the petrol station and start filling up their tanks. This causes a queue, which starts everyone else panicking. I don't know what they'll do when it runs out totally.

Warm-Up and Exam Questions

Give these questions your best shot. If they highlight areas where your knowledge falls short, it's time to re-revise those sections so you can boost your confidence for the exam.

Warm-Up Questions

1) What does a mixture consist of?
2) What are hydrocarbons made from?
3) Name three fractions obtained from crude oil.
4) How does the number of carbon atoms affect the viscosity of a hydrocarbon?
5) List three modern-day activities that depend on crude oil and its fractions.

Exam Questions

1   Crude oil can be separated into a number of different compounds as shown in the diagram:

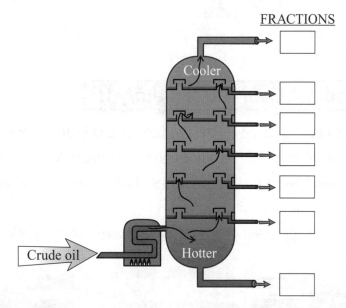

FRACTIONS

(a) (i)   Put an **M** in the box of the fraction with the longest hydrocarbon molecules.

(1 mark)

(ii) Put a **B** in the box of the fraction with the lowest boiling point.

(1 mark)

(b) Briefly explain how the separation process works.

(3 marks)

2   Even though there are many environmental problems caused by using crude oil fractions, we continue to use them mainly because

A   they are a renewable resource

B   technology is always improving

C   they are a readily available and reliable energy source

D   global warming is only a theory

(1 mark)

Cracking Crude Oil

After the distillation of crude oil (see page 55), you've still got both short and long hydrocarbons, just not all mixed together. But there's <u>more demand</u> for some products, like <u>petrol</u>, than for others.

Cracking Means Splitting Up Long–Chain Hydrocarbons...

1) <u>Long-chain hydrocarbons</u> form <u>thick gloopy liquids</u> like <u>tar</u> which aren't all that useful, so...

2) ... a lot of the longer molecules produced from <u>fractional distillation</u> are <u>turned into smaller ones</u> by a process called <u>cracking</u>.

3) Some of the products of cracking are useful as fuels, e.g. petrol for cars and paraffin for jet fuel.

4) Cracking also produces substances like <u>ethene</u>, which are needed for <u>making plastics</u> (see p.63).

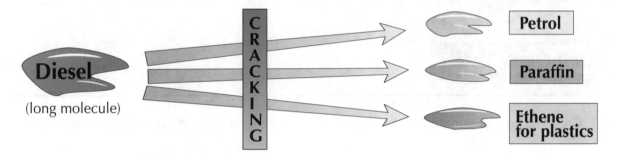

...By Passing Vapour Over a Hot Catalyst

1) <u>Cracking</u> is a <u>thermal decomposition</u> reaction — <u>breaking molecules down</u> by <u>heating</u> them.

2) The first step is to <u>heat</u> the long-chain hydrocarbon to <u>vaporise</u> it (turn it into a gas).

3) Then the <u>vapour</u> is passed over a <u>powdered catalyst</u> at a temperature of about <u>400 °C – 700 °C</u>.

4) <u>Aluminium oxide</u> is the catalyst used.

5) The <u>long-chain</u> molecules <u>split apart</u> or "crack" on the <u>surface</u> of the specks of catalyst.

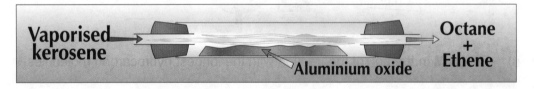

6) Most of the <u>products</u> of cracking are <u>alkanes</u> (see page 61) and unsaturated hydrocarbons called <u>alkenes</u> (see page 62)...

An alternative way of cracking long-chain hydrocarbons is to mix the vapour with steam at a very high temperature.

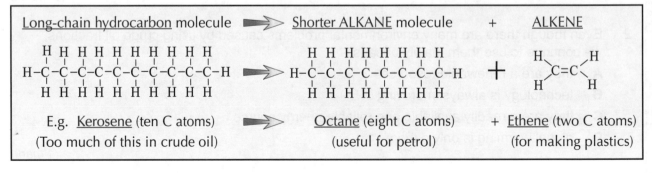

Alkanes

The <u>different fractions</u> of crude oil have <u>different structures</u>, but they're all hydrocarbons — remember, this means they're made up of just <u>carbon</u> and <u>hydrogen</u> atoms.

Crude Oil is Mostly Alkanes

1) All the fractions of crude oil are hydrocarbons called <u>alkanes</u>.

2) Alkanes are made up of <u>chains of carbon atoms</u> surrounded by <u>hydrogen atoms</u>.

$$\begin{array}{cccc} H & H & H & H \\ | & | & | & | \\ H-C-C-C-C-H \\ | & | & | & | \\ H & H & H & H \end{array}$$

3) Different alkanes have chains of different <u>lengths</u>.

4) The first four alkanes are <u>methane</u> (natural gas), <u>ethane</u>, <u>propane</u> and <u>butane</u>.

1) *Methane*

<u>Formula</u>: CH_4

(natural gas)

$$\begin{array}{c} H \\ | \\ H-C-H \\ | \\ H \end{array}$$

2) *Ethane*

<u>Formula</u>: C_2H_6

$$\begin{array}{cc} H & H \\ | & | \\ H-C-C-H \\ | & | \\ H & H \end{array}$$

3) *Propane*

<u>Formula</u>: C_3H_8

$$\begin{array}{ccc} H & H & H \\ | & | & | \\ H-C-C-C-H \\ | & | & | \\ H & H & H \end{array}$$

4) *Butane*

<u>Formula</u>: C_4H_{10}

$$\begin{array}{cccc} H & H & H & H \\ | & | & | & | \\ H-C-C-C-C-H \\ | & | & | & | \\ H & H & H & H \end{array}$$

Each straight line shows a covalent bond (page 19).

5) Carbon atoms form <u>four bonds</u> and hydrogen atoms only form <u>one bond</u>. The diagrams above show that all the atoms have formed bonds with as many other atoms as they can — this means they're <u>saturated</u>.

6) Alkanes all have the <u>general formula</u> C_nH_{2n+2}. So if an alkane has 5 carbons, it's got to have $(2 \times 5) + 2 = 12$ hydrogens.

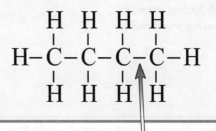

Alkanes
$= C_nH_{2n+2}$

Alkenes

Alkenes are hydrocarbons too. Here are a few basics about their structure and how to test for them.

Alkenes Have a C=C Double Bond

1) Alkenes are hydrocarbons which have a double bond between two of the carbon atoms in their chain.

2) They are known as unsaturated because they can make more bonds — the double bond can open up, allowing the two carbon atoms to bond with other atoms.

3) The first two alkenes are ethene (with two carbon atoms) and propene (three Cs).

4) All alkenes have the general formula: C_nH_{2n} — they have twice as many hydrogens as carbons.

1) Ethene

Formula: C_2H_4

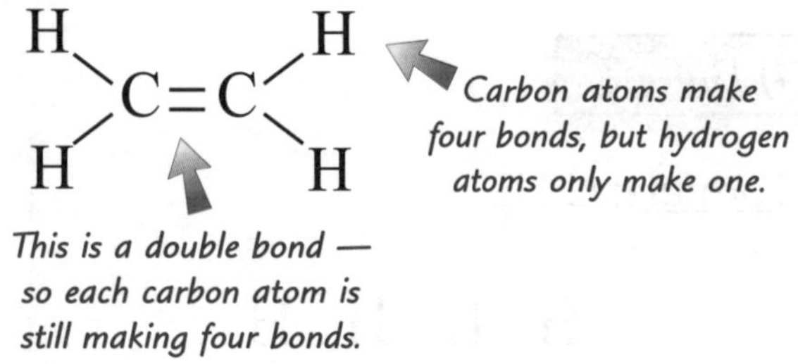

Carbon atoms make four bonds, but hydrogen atoms only make one.

This is a double bond — so each carbon atom is still making four bonds.

2) Propene

Formula: C_3H_6

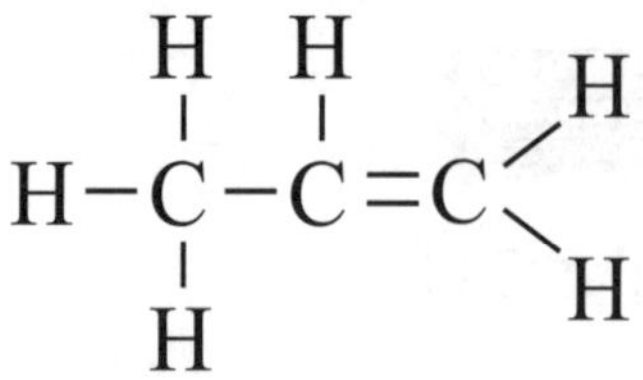

Alkenes Turn Bromine Water Colourless

1) You can test for an alkene by adding the substance to bromine water.

2) An alkene will decolourise the bromine water, turning it from orange to colourless.

3) This is because the double bond has opened up and formed bonds with the bromine.

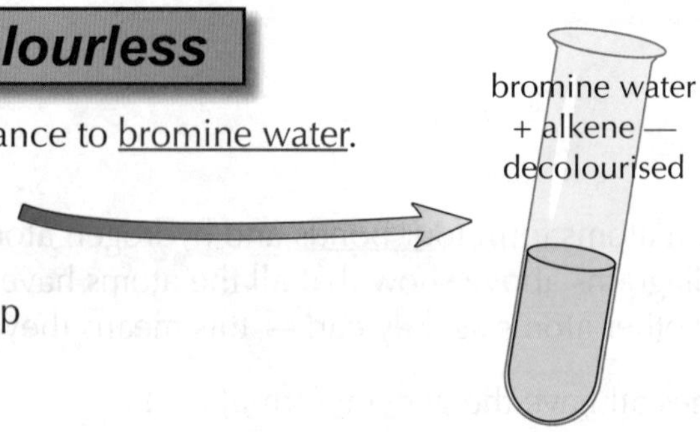

bromine water
+ alkene —
decolourised

That double bond makes all the difference...

Don't get alkenes confused with alkanes. Alkenes have a C=C bond, alkanes don't. The first part of their names is the same though. "Meth-" means "one carbon atom", "eth-" means "two C atoms", "prop-" means "three C atoms", "but-" means "four C atoms", etc.

Using Alkenes to Make Polymers

Don't get stuck into the exciting world of <u>polymers</u> until you're up to scratch with the <u>alkene</u> stuff on the previous page — the knowledge will come in useful here...

Alkenes *Can Be Used to Make* Polymers

1) Probably the most useful thing you can do with alkenes is <u>polymerisation</u>.

2) This means joining together lots of <u>small alkene molecules</u> (<u>monomers</u>) to form <u>very large molecules</u> — these long-chain molecules are called <u>polymers</u>.

Polymers are often written without the brackets — e.g. polyethene.

3) For instance, many <u>ethene</u> molecules can be joined up to produce <u>poly(ethene)</u> or "polythene".

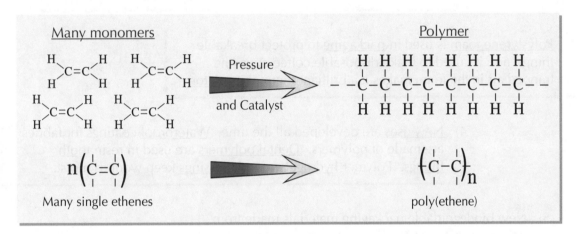

4) In the same way, if you join lots of <u>propene</u> molecules together, you've got <u>poly(propene)</u>.

Different Polymers *Have Different* Physical Properties

1) The physical properties of a polymer depend on <u>what it's made from</u>. Polyamides are usually stronger than poly(ethene), for example.

2) A polymer's <u>physical properties</u> are also affected by the <u>temperature and pressure</u> of polymerisation.

3) Poly(ethene) made at <u>200 °C</u> and <u>2000 atmospheres pressure</u> is <u>flexible</u>, and has <u>low density</u>. But poly(ethene) made at <u>60 °C</u> and a <u>few atmospheres pressure</u> with a <u>catalyst</u> is <u>rigid</u> and <u>dense</u>.

Polymers are really important...

...and now you know the basics of how they're made. Monomers are often alkenes that contain double bonds. When they are put under <u>pressure</u> in the presence of a <u>catalyst</u>, they join together to form really big, long-chain molecules called polymers. So remember <u>monomers</u> form <u>polymers</u>...

Polymers

Polymers Are Suitable For Various Different **Uses**

1) <u>Light, stretchable</u> polymers such as low density poly(ethene) are used to make plastic bags. <u>Elastic</u> polymer fibres are used to make super-stretchy <u>LYCRA® fibre</u> for tights.

2) <u>PVC</u> is <u>strong</u> and durable, and it can be made either <u>rigid</u> or <u>stretchy</u>. The rigid kind is used to make <u>window frames and piping</u>. The stretchy kind is used to make <u>synthetic leather</u>.

3) <u>Polystyrene foam</u> is used in <u>packaging</u> to protect breakable things, and it's used to make disposable coffee cups (the trapped air in the foam makes it a brilliant <u>thermal insulator</u>).

4) <u>New uses</u> are developed all the time. <u>Waterproof</u> coatings for fabrics are made of polymers. <u>Dental polymers</u> are used in resin <u>tooth fillings</u>. Polymer <u>hydrogel wound dressings</u> keep wounds moist.

5) <u>New biodegradable packaging</u> materials made from polymers and <u>cornstarch</u> are being produced.

6) <u>Memory foam</u> is an example of a <u>smart material</u>. It's a polymer that gets <u>softer</u> as it gets <u>warmer</u>. Mattresses can be made of memory foam — they mould to your body shape when you lie on them.

Polymers Are Cheap, But Most **Don't Rot**

1) Most polymers aren't "<u>biodegradable</u>" — they're not broken down by microorganisms, so they <u>don't rot</u>.

2) It's difficult to get rid of them — if you bury them in a landfill site, they'll <u>still</u> be there <u>years later</u>. The best thing is to <u>re-use</u> them as many times as possible and then <u>recycle</u> them if you can.

3) Things made from polymers are usually <u>cheaper</u> than things made from metal. However, as <u>crude oil resources</u> get <u>used up,</u> the <u>price</u> of crude oil will rise. Crude oil products like polymers will get dearer.

4) It may be that one day there won't be <u>enough</u> oil for fuel AND plastics AND all the other uses. Choosing how to use the oil that's left means weighing up advantages and disadvantages on all sides.

Revision's like a polymer — you join lots of little facts up...

Polymers are all over the place — and I don't just mean all those plastic bags stuck in trees. There are naturally occurring polymers, like <u>rubber</u> and <u>silk</u>. That's quite a few clothing options, even without synthetic polymers like <u>polyester</u> and <u>PVC</u>. You also have polymers on the inside — <u>DNA's</u> a polymer.

Burning Fuels

We get loads of fuels from oil. And then we burn them. But there's <u>burning</u> and there's <u>burning</u>...

Complete Combustion Happens When There's Plenty of Oxygen

1) When there's <u>plenty of oxygen</u> about, hydrocarbons burn to produce only <u>carbon dioxide</u> and <u>water</u>.

> **hydrocarbon + oxygen → carbon dioxide + water**   **(+ energy)**

2) The <u>hydrogen</u> and <u>carbon</u> in the hydrocarbon have both been <u>oxidised</u>.

3) Many <u>gas room heaters</u> release these <u>waste gases</u> into the room, which is perfectly OK. As long as the gas heater is <u>working properly</u> and the room is <u>well ventilated</u>, there's no problem.

4) <u>Complete combustion</u> releases <u>lots of energy</u> and only produces those two <u>harmless waste products</u>. When there's <u>plenty of oxygen</u> and combustion is complete, the gas burns with a <u>clean blue flame</u>.

Lots of CO_2 isn't ideal, but the alternatives are worse (see below).

5) Here's the <u>equation</u> for the complete combustion of <u>methane</u> — a simple hydrocarbon fuel.

You can test for CO_2 by bubbling the gas through limewater. It turns limewater milky (see page 31).

Natural gas is mostly methane (CH_4).

> $$CH_4 + 2O_2 \rightarrow 2H_2O + CO_2$$

Partial Combustion of Hydrocarbons is NOT Safe

1) If there <u>isn't enough oxygen</u> the combustion will be <u>partial</u>. Carbon dioxide and water are still produced, but you can also get <u>carbon monoxide</u> (CO) and <u>carbon</u>.

2) Partial combustion means a <u>smoky yellow flame</u>, and <u>less energy</u> than complete combustion.

> **hydrocarbon + oxygen → carbon + carbon monoxide + carbon dioxide + water**
>
> **(+ energy)**

3) The <u>carbon monoxide</u> is a <u>colourless</u>, <u>odourless</u> and very toxic (<u>poisonous</u>) gas.

4) Every year people are <u>killed</u> while they sleep due to <u>faulty</u> gas fires and boilers filling the room with <u>carbon monoxide</u> and nobody realising — this is why it's important to <u>regularly service gas appliances</u>. The black carbon given off produces <u>sooty marks</u> — a <u>clue</u> that the fuel is <u>not</u> burning fully.

5) So basically, you want <u>lots of oxygen</u> when you're burning fuel — you get <u>more energy</u> given out, and you don't get any <u>messy soot</u> or <u>poisonous gases</u>.

6) Here's an example of an <u>equation</u> for partial combustion too.

> $$4CH_4 + 6O_2 \rightarrow C + 2CO + CO_2 + 8H_2O$$

This is just one possibility. The products depend on how much oxygen is present. E.g. you could also have: $4CH_4 + 7O_2 \rightarrow 2CO + 2CO_2 + 8H_2O$ — the important thing is that the equation is balanced (see p. 23).

There's Lots to Consider When Choosing the Best Fuel

1) <u>Ease of ignition</u> — whether it <u>burns easily</u>. Fuels like gas burn more easily than diesel.

2) <u>Energy value</u> — the amount of energy released.

3) <u>Ash and smoke</u> — some fuels, like coal, leave behind a lot of <u>ash</u> that needs to be disposed of.

4) <u>Storage and transport</u> — gas needs to be stored in special <u>canisters</u> and coal needs to be kept <u>dry</u>. Fuels need to be transported <u>carefully</u> as gas leaks and oil spills can be dangerous.

Warm-Up and Exam Questions

Time for some more questions now. Work your way through them, then go back and look again at anything that you struggled with. It's the only way that you'll get all of this stuff into your head...

Warm-Up Questions

1) Describe the conditions used for cracking hydrocarbons.
2) Name the first three alkanes.
3) Why are alkenes described as unsaturated hydrocarbons?
4) What are the problems with disposing of polymers?
5) What kind of combustion occurs when there's plenty of oxygen?

Exam Questions

1 Alkanes are made up of chains of carbon atoms surrounded by hydrogen atoms.

   (a) Butane contains four carbon atoms. Give its formula.

(1 mark)

   (b) Draw the displayed formula of butane showing all of the bonds.

(1 mark)

2 The bonds in alkanes are best described as

   A   carbon-carbon single bonds and carbon-hydrogen single bonds

   B   carbon-carbon single bonds and carbon-hydrogen double bonds

   C   carbon-carbon double bonds and carbon-hydrogen single bonds

   D   carbon-carbon double bonds and carbon-hydrogen double bonds

(1 mark)

3 A test for an alkene is

   A   universal indicator is decolourised

   B   starch solution goes dark blue

   C   bromine water is decolourised

   D   bromine water remains brown

(1 mark)

4 Polymers have many different uses.

   (a) Give two different uses for polymers.

(2 marks)

   (b) Explain one disadvantage of using polymers in industry.

(1 mark)

Exam Questions

5   (a) Complete this equation for the formation of polypropene.

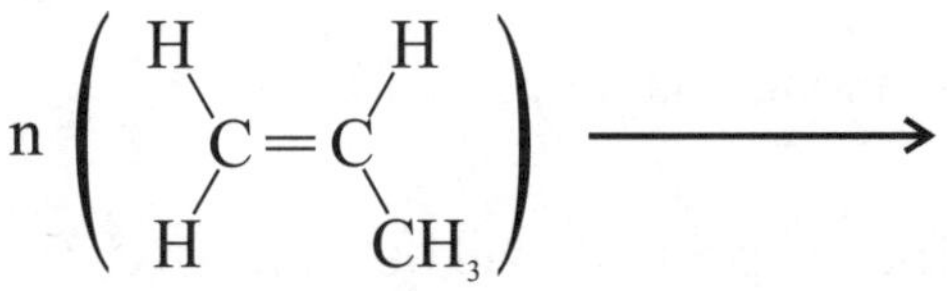

(1 mark)

     (b) The structural formula of polystyrene is shown below.

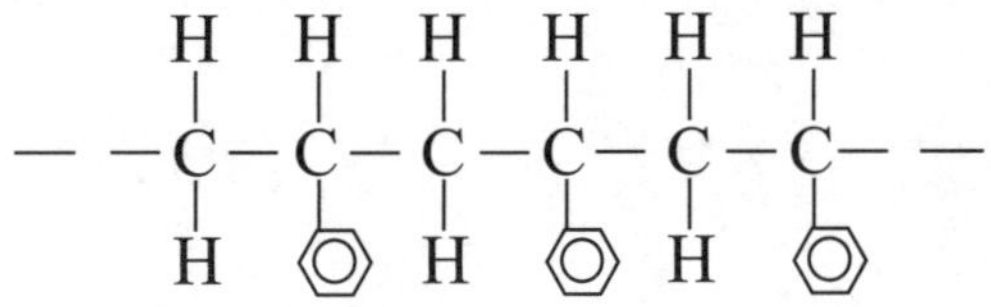

         Draw the structural formula of its monomer.

(1 mark)

6   (a) Describe the conditions necessary for complete combustion and
        partial combustion to occur.

(2 marks)

     (b) Which of these are the products of complete combustion?

        A   carbon monoxide and water

        B   carbon dioxide and water

        C   water and oxygen

        D   sulfur dioxide and water

(1 mark)

     (c) Explain why partial combustion could be dangerous.

(2 marks)

7   Many hydrocarbons produced from the fractional distillation of crude oil are cracked.

     (a) Explain why hydrocarbons are cracked.

(1 mark)

     (b) Describe the process used to crack the hydrocarbons found in crude oil.

(3 marks)

Revision Summary for Section 3

Some people skip these pages. But what's the point in reading that great big section if you're not going to check if you really know it or not. Look, just read the first ten questions, and I guarantee there'll be an answer you'll have to look up. And when it comes up in the exam, you'll be so glad you did.

1) What does crude oil consist of? What does fractional distillation do to crude oil?

2) Is a short-chain hydrocarbon more viscous than a long-chain hydrocarbon?
   Is it more volatile?

3)* You're going on holiday to a very cold place. The temperature will be about –10 °C.
   Which of the fuels shown below do you think will work best in your camping stove?
   Explain your answer.

Fuel	Boiling point (°C)
Propane	–42
Butane	–0.4
Pentane	36.2

4) Name the two important types of bond in crude oil.

5) Why do big hydrocarbon molecules have higher boiling points than small ones?

6) What is "cracking"?

7) Give a typical example of a substance that is cracked, and the products that you get from cracking it.

8) What's the general formula for an alkane?

9) What kind of carbon-carbon bond do alkenes have?

10) What is the general formula for alkenes?

11) Draw the displayed formula of ethene.

12) What are polymers? What kinds of substances can form polymers?

13) Give two factors which affect the physical properties of a polymer.

14) Why might polymers become more expensive in the future?

15) Give two reasons why complete combustion is better than partial combustion when you're burning fuel to heat a house.

16) Give four things you might want to consider when deciding on the best fuel to use.

* Answers on page 253

Plant Oils

Plant oils come from <u>plants</u>. I know it's tricky, but just do your best to remember.

We can **extract oils** from **plants**

olive mush

weight

olive oil

1) Some <u>fruits</u> and <u>seeds</u> contain a lot of <u>oil</u>. For example, avocados and olives are oily fruits. Brazil nuts, peanuts and sesame seeds are oily seeds (a nut is just a big seed really).

2) These oils can be extracted and used for <u>food</u> or for <u>fuel</u>.

3) To get the oil out, the plant material is <u>crushed</u>. The next step is to <u>press</u> the crushed plant material between metal plates and squash the oil out. This is the traditional method of producing <u>olive oil</u>.

4) Oil can be separated from crushed plant material by a <u>centrifuge</u> — rather like using a spin-dryer to get water out of wet clothes.

5) Or <u>solvents</u> can be used to get oil from plant material.

6) <u>Distillation</u> refines oil, and <u>removes water</u>, <u>solvents</u> and <u>impurities</u>.

Vegetable oils are used in food

1) Vegetable oils provide a lot of <u>energy</u> — they have a very high energy content.

2) There are other nutrients in vegetable oils. For example, oils from seeds contain <u>vitamin E</u>.

3) Vegetable oils contain <u>essential fatty acids</u>, which the body needs for many metabolic processes.

Vegetable oils have benefits for cooking

1) Vegetable oils have <u>higher boiling points</u> than water. This means they can be used to cook foods at higher temperatures and at <u>faster</u> speeds.

2) Cooking with vegetable oil gives food a <u>different flavour</u>. This is because of the oil's <u>own</u> flavour, but it's also down to the fact that many flavours come from chemicals that are <u>soluble</u> in oil. This means the oil '<u>carries</u>' the flavour, making it seem more <u>intense</u>.

3) Using oil to cook food <u>increases</u> the <u>energy</u> we get from eating it.

Vegetable oils can be used to produce fuels

1) Vegetable oils such as rapeseed oil and soybean oil can be <u>processed</u> and turned into <u>fuels</u>.

2) Because vegetable oils provide a lot of <u>energy</u> they're really suitable for use as fuels.

3) A particularly useful fuel made from vegetable oils is called <u>biodiesel</u>. Biodiesel has similar properties to ordinary diesel fuel — it burns in the same way, so you can use it to fuel a diesel engine.

See page 94 for more about biodiesel.

Plant Oils

Oils are usually quite runny at room temperature. That's fine for salad dressing, say, but not so good for spreading on your sandwiches. For that, you could <u>hydrogenate</u> the oil to make <u>margarine</u>...

Unsaturated oils contain C=C double bonds

1) Oils and fats contain <u>long-chain molecules</u> with lots of <u>carbon</u> atoms.

2) Oils and fats are either <u>saturated</u> or <u>unsaturated</u>.

3) Unsaturated oils contain <u>double bonds</u> between some of the carbon atoms in their carbon chains.

4) So, an unsaturated oil will <u>decolourise</u> bromine water (as the bromine opens up the double bond and joins on).

5) <u>Monounsaturated</u> fats contain <u>one</u> C=C double bond somewhere in their carbon chains. <u>Polyunsaturated</u> fats contain <u>more than one</u> C=C double bond.

bromine water
+ unsaturated oil
— decolourised

Unsaturated oils can be hydrogenated

1) <u>Unsaturated</u> vegetable oils are <u>liquid</u> at room temperature.

2) They can be hardened by reacting them with <u>hydrogen</u> in the presence of a <u>nickel catalyst</u> at about <u>60 °C</u>. This is called <u>hydrogenation</u>. The hydrogen reacts with the double-bonded carbons and opens out the double bonds.

$$C=C + H_2 \xrightarrow[\text{catalyst}]{\text{nickel}} H-C-C-H$$

3) Hydrogenated oils have <u>higher melting points</u> than unsaturated oils, so they're <u>more solid</u> at room temperature. This makes them useful as <u>spreads</u> and for baking cakes and pastries.

4) Margarine is usually made from <u>partially</u> hydrogenated vegetable oil — turning <u>all</u> the double bonds in vegetable oil to single bonds would make margarine <u>too hard</u> and difficult to spread. Hydrogenating <u>most</u> of them gives margarine a nice, buttery, spreadable consistency.

5) Partially hydrogenated vegetable oils are often used instead of butter in processed foods, e.g. biscuits. These oils are a lot <u>cheaper</u> than butter and they <u>keep longer</u>. This makes biscuits cheaper and gives them a long shelf life.

6) But partially hydrogenating vegetable oils means you end up with a lot of so-called <u>trans fats</u>. And there's evidence to suggest that trans fats are <u>very bad</u> for you.

Vegetable oils in foods can affect health

1) Vegetable oils tend to be <u>unsaturated</u>, while animal fats tend to be <u>saturated</u>.

2) In general, <u>saturated fats</u> are less healthy than <u>unsaturated fats</u> (as <u>saturated</u> fats <u>increase</u> the amount of <u>cholesterol</u> in the blood, which can block up the arteries and increase the risk of <u>heart disease</u>).

3) Natural <u>unsaturated</u> fats such as olive oil and sunflower oil <u>reduce</u> the amount of blood cholesterol. But because of the trans fats, <u>partially hydrogenated vegetable oil</u> increases the amount of <u>cholesterol</u> in the blood. So eating a lot of foods made with partially hydrogenated vegetable oils can actually increase the risk of heart disease.

4) <u>Cooking</u> food in oil, whether saturated, unsaturated or partially hydrogenated, makes it more <u>fattening</u>.

Emulsions

Emulsions are all over the place in <u>foods</u>, <u>cosmetics</u> and <u>paint</u>.

Emulsions can be made from oil and water

1) Oils <u>don't dissolve in water</u>. So far so good...

2) However, you <u>can</u> mix an oil with water to make an <u>emulsion</u>. Emulsions are made up of lots of <u>droplets</u> of one liquid <u>suspended</u> in another liquid. You can have an oil-in-water emulsion (oil droplets suspended in water) or a water-in-oil emulsion (water droplets suspended in oil).

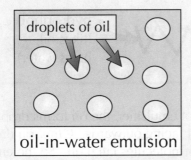

oil-in-water emulsion

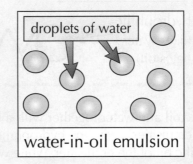

water-in-oil emulsion

3) Emulsions are <u>thicker</u> than either oil or water. E.g. mayonnaise is an emulsion of sunflower oil (or olive oil) and vinegar — it's thicker than either.

4) The physical properties of emulsions make them suited to <u>lots of uses</u> in food — e.g. as salad dressings and in sauces. For instance, a salad dressing made by shaking olive oil and vinegar together forms an <u>emulsion</u> that <u>coats</u> salad better than plain oil or plain vinegar.

5) Generally, the <u>more oil</u> you've got in an oil-in-water emulsion, the <u>thicker</u> it is. Milk is an oil-in-water emulsion with not much oil and a lot of water — there's about 3% oil in full-fat milk. Single cream has a bit more oil — about 18%. Double cream has lots of oil — nearly 50%.

6) <u>Whipped cream</u> and ice cream are oil-in-water emulsions with an extra ingredient — <u>air</u>. Air is whipped into cream to give it a <u>fluffy</u>, frothy consistency for use as a topping. Whipping air into ice cream gives it a <u>softer texture</u>, which makes it easier to scoop out of the tub.

7) Emulsions also have <u>non-food uses</u>. Most <u>moisturising lotions</u> are oil-in-water emulsions. The smooth texture of an emulsion makes it easy to rub into the skin.

Emulsions

Some foods contain **emulsifiers** to help **oil** and **water mix**

Oil and water mixtures naturally <u>separate out</u>. But here's where emulsifiers come in...

1) Emulsifiers are molecules with one part that's <u>attracted to water</u> and another part that's <u>attracted to oil</u> or fat. The bit that's attracted to water is called <u>hydrophilic</u>, and the bit that's attracted to oil is called <u>hydrophobic</u>.

2) The <u>hydrophilic</u> end of each emulsifier molecule latches onto <u>water molecules</u>.

3) The <u>hydrophobic</u> end of each emulsifier molecule cosies up to <u>oil molecules</u>.

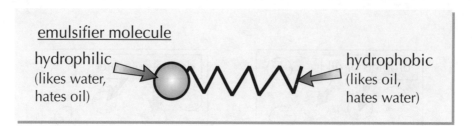

emulsifier molecule

hydrophilic
(likes water,
hates oil)

hydrophobic
(likes oil,
hates water)

4) When you shake oil and water together with a bit of emulsifier, the oil forms droplets, surrounded by a coating of emulsifier... <u>with the hydrophilic bit facing outwards</u>. Other oil droplets are <u>repelled</u> by the hydrophilic bit of the emulsifier, while water molecules latch on. So the emulsion won't separate out. Clever.

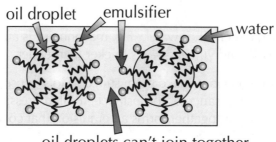

oil droplet   emulsifier   water

oil droplets can't join together

Using **emulsifiers** has **pros** and **cons**

1) Emulsifiers <u>stop</u> emulsions from <u>separating</u> out and this gives them a longer <u>shelf-life</u>.

2) Emulsifiers allow food companies to produce food that's <u>lower in fat</u> but that still has a <u>good texture</u>.

3) The <u>down side</u> is that some people are <u>allergic</u> to certain emulsifiers. For example, <u>egg yolk</u> is often used as an emulsifier — so people who are allergic to eggs need to <u>check</u> the <u>ingredients</u> very carefully.

Emulsion paint — spread mayonnaise all over the walls...

Before fancy stuff from abroad like olive oil, we fried our bacon and eggs in <u>lard</u>. Lard wouldn't be so good for making salad cream though. Emulsions like salad cream have to be made from shaking up two liquids — tiny droplets of one liquid are 'suspended' (NOT dissolved) in the other liquid.

Warm-Up and Exam Questions

Here we go again... Yet another lovely page of questions to test your mad chemistry skills. Start with the warm-up questions and once you're happy you've nailed 'em have a crack at the exam questions.

Warm-Up Questions

1) Why are vegetable oils suitable for use as fuels?
2) What conditions are used for the hydrogenation of unsaturated vegetable oils?
3) Why are natural unsaturated fats typically healthier than saturated fats?
4) What is an emulsion?

Exam Questions

1   Oils that are extracted from plants and refined are often used for cooking.

   (a)   Describe the process used to extract and refine oils from plants.

(4 marks)

   (b)   Give **three** benefits of using vegetable oils for cooking.

(3 marks)

2   Mayonnaise contains emulsifiers — molecules that have a hydrophilic end and a hydrophobic end.

   (a)   On this diagram of an emulsifier molecule, label the hydrophilic and hydrophobic ends.

(1 mark)

   (b)   Explain the meanings of hydrophilic and hydrophobic.

(2 marks)

   (c)   The diagram below shows an oil molecule in water.
         Show on the diagram how emulsifier molecules arrange themselves.

(1 mark)

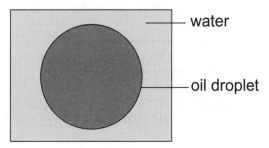

   (d)   What effect do emulsifiers have on the mayonnaise?

(1 mark)

Alcohols

This page is about different types of <u>alcohols</u> — and that's not just beer, wine and spirits.

Alcohols have an '-OH' functional group and end in '-ol'

1) The <u>general formula</u> of an alcohol is $C_nH_{2n+1}OH$.
So an alcohol with 2 carbons has the formula C_2H_5OH.

A homologous series is a group of chemicals that react in a similar way because they have the same functional group (in alcohols it's the −OH group).

2) All alcohols contain the same <u>-OH group</u>.
You need to know the <u>first 3</u> in the homologous series:

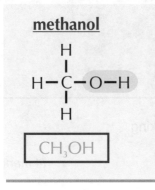

methanol

CH_3OH

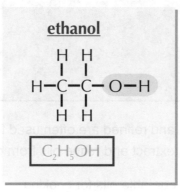

ethanol

C_2H_5OH

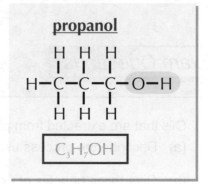

propanol

C_3H_7OH

3) The basic <u>naming</u> system is the same as for alkanes — but replace the final '<u>-e</u>' with '<u>-ol</u>'.

4) Don't write CH_4O instead of CH_3OH — it doesn't show the <u>functional -OH group</u>.

The first three alcohols have similar properties

1) Alcohols are <u>flammable</u>. They burn in air to produce <u>carbon dioxide</u> and <u>water</u>.
E.g.

$$2CH_3OH_{(l)} + 3O_{2(g)} \rightarrow 2CO_{2(g)} + 4H_2O_{(g)}$$

2) The first three alcohols all <u>dissolve completely in water</u> to form <u>neutral solutions</u>.

3) They also react with <u>sodium</u> to give <u>hydrogen</u> and <u>alkoxides</u>,
e.g. ethanol gives sodium ethoxide and H_2.
E.g.

$$2C_2H_5OH_{(l)} + 2Na_{(s)} \rightarrow 2C_2H_5ONa_{(aq)} + H_{2(g)}$$

4) <u>Ethanol</u> is the main alcohol in alcoholic drinks. It's not as <u>toxic</u> as methanol (which causes <u>blindness</u> if drunk) but it still damages the <u>liver</u> and <u>brain</u>.

All alcohols have the functional group '-OH'

Methanol, ethanol and propanol are the first three alcohols in the alcohol homologous series. It's not the catchiest set of names in the world, I grant you, but sometimes that's just the way life goes...

Alcohols

Here's another page of useful facts about alcohols.

Alcohols are used as solvents

1) Alcohols such as methanol and ethanol can <u>dissolve</u> most compounds that <u>water</u> dissolves, but they can also dissolve substances that <u>water can't dissolve</u> — e.g. hydrocarbons, oils and fats. This makes ethanol, methanol and propanol <u>very useful solvents</u> in industry.

2) <u>Ethanol</u> is the solvent for <u>perfumes</u> and <u>aftershave</u> lotions.
It can mix with both the <u>oils</u> (which give the smell) <u>and</u> the <u>water</u> (that makes up the bulk).

3) '<u>Methylated spirit</u>' (or 'meths') is <u>ethanol</u> with chemicals (e.g. methanol) added to it. It's used to <u>clean</u> paint brushes and as a <u>fuel</u> (among other things).
It's <u>poisonous</u> to drink, so a <u>purply-blue dye</u> is also added (to stop people drinking it by mistake).

Alcohols are used as fuels

1) Ethanol is used as a fuel in <u>spirit burners</u> — it burns fairly cleanly and it's non-smelly.

2) Ethanol can also be mixed in with petrol and used as <u>fuel for cars</u>. Since pure ethanol is <u>clean burning</u>, the more ethanol in a petrol/ethanol mix, the less <u>pollution</u> is produced.

3) Some countries that have little or no oil deposits but plenty of land and sunshine (e.g. Brazil) grow loads of <u>sugar cane</u>, which they <u>ferment</u> to form ethanol.

4) A big advantage of this is that sugar cane is a <u>renewable resource</u> (unlike petrol, which will run out).

Quick tip — don't fill your car with single malt whisky...

This page is stuffed full of facts about the uses of alcohols. Once you've read over them, write down everything you've learnt and see how much you know. You may even surprise yourself...

Carboxylic Acids

So what if carboxylic is a funny name — these are easy.

Carboxylic acids have the functional group *-COOH*

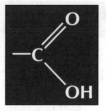

1) <u>Carboxylic acids</u> have '-COOH' as a <u>functional group</u>.

2) Their names end in '-<u>anoic acid</u>' (and start with the normal '<u>meth</u>/<u>eth</u>/<u>prop</u>').

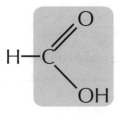 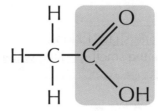

Methanoic acid
HCOOH

Ethanoic acid
CH₃COOH

Propanoic acid
C₂H₅COOH

Carboxylic acids react with *Carbonates*

1) They react to produce <u>carbon dioxide</u>.
There's more about reactions of acids on page 149.

2) The <u>salts</u> formed in these reactions end in -<u>anoate</u> — e.g. methanoic acid will form a <u>methanoate</u>, ethanoic acid an <u>ethanoate</u>, etc. For example:

> ethanoic acid + sodium carbonate → carbon dioxide + sodium ethanoate

3) Carboxylic acids <u>dissolve in water</u> to produce <u>acidic solutions</u>. When they dissolve, they <u>ionise</u> and <u>release H⁺ ions</u> which are responsible for making the solution <u>acidic</u>. But, because they <u>don't ionise completely</u> (not many H⁺ ions are released), they just form <u>weak acidic solutions</u>. This means that they have a <u>higher pH</u> (less acidic) than aqueous solutions of <u>strong acids</u> with the <u>same concentration</u>.

The strength of an acid isn't the same as its concentration (see page 147). Concentration is how watered down your acid is and strength is how well it has ionised in water.

Carboxylic acids are just like other acids

The trickiest bit on this page is probably the bit about carboxylic acids <u>not ionising completely</u> in water and being <u>weak acids</u>. But when it comes to carbonates, they just act like any old acid — easy.

Carboxylic Acids

Here are some of the wonderful things you can do with carboxylic acids.

Some *carboxylic acids* are fairly *common*

1) <u>Ethanoic acid</u> can be made by <u>oxidising ethanol</u>. Microbes, like yeast, cause the ethanol to ferment. Ethanol can also be oxidised using <u>oxidising agents</u>.

ethanol + oxygen → ethanoic acid + water

If you leave wine open, the ethanol in it is oxidised — this is why it goes off.

2) <u>Ethanoic acid</u> can then be <u>dissolved in water</u> to make <u>vinegar</u>, which is used for <u>flavouring</u> and <u>preserving</u> foods.

3) <u>Citric acid</u> (another carboxylic acid) is present in <u>oranges</u> and <u>lemons</u>, and is manufactured in large quantities to make <u>fizzy drinks</u>. It's also used to get rid of <u>scale</u> (see page 207).

Carboxylic acids are used in industry to make *soaps* and *esters*

1) <u>Carboxylic acids</u> with <u>longer chains</u> of carbon atoms are used to make <u>soaps and detergents</u>.

2) Carboxylic acids are also used in the preparation of <u>esters</u> (see next page).

3) Ethanoic acid is a very good <u>solvent</u> for many organic molecules. But ethanoic acid isn't usually chosen as a solvent because it makes the solution <u>acidic</u>.

Ethanoic acid — it's not just for putting on your chips...

Bet you didn't think of <u>vinegar</u> or <u>fizzy drinks</u> when you first heard of carboxylic acids. It turns out they've got a fair few uses from <u>descaler</u> to <u>soaps</u> — which is pretty handy really.

Esters

Mix an alcohol from page 74 and a carboxylic acid from page 76, and you get an ester.

Esters have the functional group -COO-

1) Esters are formed from an alcohol and a carboxylic acid.
2) An acid catalyst is usually used (e.g. concentrated sulfuric acid).

$$alcohol + carboxylic\ acid \rightarrow ester + water$$

| CH₃COOH | C₂H₅OH | CH₃COOC₂H₅ | H₂O |
| ethanoic acid | ethanol | ethyl ethanoate | water |

Their names end in '-oate'. The alcohol forms the first part of the ester's name, and the acid forms the second part.

ethanol + ethanoic acid → ethyl ethanoate + water

methanol + propanoic acid → methyl propanoate + water

Esters smell nice but don't mix well with water

1) Many esters have pleasant smells — often quite sweet and fruity. They're also volatile. This makes them ideal for perfumes (the evaporated molecules can be detected by smell receptors in your nose).
2) However, many esters are flammable (or even highly flammable). So their volatility also makes them potentially dangerous.
3) Esters don't mix very well with water. (They're not nearly as soluble as alcohols or carboxylic acids.)
4) But esters do mix well with alcohols and other organic solvents.

Esters are often used in flavourings and perfumes

1) Because many esters smell nice, they're used in perfumes.
2) Esters are also used to make flavourings and aromas — e.g. there are esters that smell or taste of rum, apple, orange, banana, grape, pineapple, etc.
3) Some esters are used in ointments (they give Deep Heat® its smell).
4) Other esters are used as solvents for paint, ink, glue and in nail varnish remover.

There are things you need to think about when using esters:

1) Inhaling the fumes from some esters irritates mucous membranes in the nose and mouth.
2) Ester fumes are heavier than air and very flammable. Flammable vapour + naked flame = flash fire.
3) Some esters are toxic, especially in large doses. Some people worry about health problems associated with synthetic food additives such as esters.
4) BUT... esters aren't as volatile or as toxic as some other organic solvents — they don't release nearly as many toxic fumes as some of them. In fact esters have replaced solvents such as toluene in many paints and varnishes.

Warm-Up and Exam Questions

Almost done with this section. Just a few more questions to get stuck into.

Warm-Up Questions

1) What is the general formula of an alcohol?
2) Give two uses of methylated spirit.
3) Name the first three carboxylic acids in the homologous series.
4) Give the functional group of an ester.

Exam Questions

1   Carboxylic acids are a widely used family of organic chemicals.

(a) Ethanoic acid is better known by the name of its dilute solution — vinegar.
Draw its displayed formula.
(2 marks)

(b) One use of carboxylic acids is in the production of esters.

(i)   Name the ester formed when ethanoic acid is reacted with ethanol.
(1 mark)

(ii) When carboxylic acids and alcohols react to form esters,
one other product is formed. Name this product.
(1 mark)

(iii) Give one use of esters.
(1 mark)

(c) Carboxylic acids are weak acids. Explain what this means.
(2 marks)

2   Alcohols are an important group of organic chemicals.
The most widely used alcohol is ethanol. Its displayed formula is shown below.

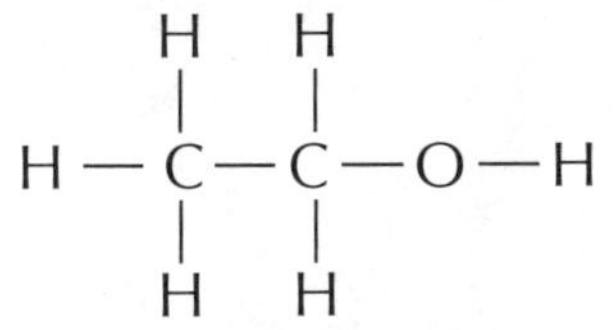

```
      H   H
      |   |
H  —  C — C — O — H
      |   |
      H   H
```

(a) Name two other chemicals in this homologous series.
(2 marks)

(b) Write down the functional group of ethanol.
(1 mark)

(c) Ethanol can be used as a fuel.
Ethanol burns in oxygen to give carbon dioxide and water.
Write a balanced symbol equation for this reaction.
(2 marks)

(d) Ethanol reacts with sodium to give sodium ethoxide
and one other product. Name this product.
(1 mark)

(e) Ethanol can be used as a solvent.
Suggest one disadvantage of using ethanol as a solvent.
(1 mark)

Revision Summary for Section 4

Plant oils, emulsions, alcohols and carboxylic acids — can they really all belong in the same section, I almost seem to hear you ask. Yes they can is my answer, and it'll be really helpful if you know them all inside out. Try these questions and see how much you really know:

1) Why do some oils need to be distilled after they have been extracted?

2) List two advantages of cooking with oil.

3) Apart from cooking, list a use of vegetable oils.

4) What kind of carbon-carbon bond do unsaturated oils contain?

5) What happens when you react unsaturated oils with hydrogen?

6) Why do some foods contain partially hydrogenated vegetable oil instead of butter?

7) What would make an emulsion thicker? Give an example of an emulsion.

8) How do emulsifiers keep emulsions stable?

9) Suggest one problem of adding emulsifiers to food.

10) Draw the structure of the first three alcohols.

11) When alcohols dissolve in water, is the solution acidic, neutral or alkaline?

12) What gas is formed when alcohols react with sodium?

13) Give two uses of alcohols.

14) What is the functional group in carboxylic acids?

15) Give two uses of carboxylic acids.

16) What two kinds of substance react together to form an ester?
    What catalyst is used in the formation of esters?

17) Write down three properties of esters.

The Earth's Structure

No one accepted the theory of <u>plate tectonics</u> for ages. Almost everyone does now. How times change.

The **Earth** has a **Crust**, **Mantle** and **Core**

1) The <u>crust</u> is Earth's thin outer layer of solid rock (its average depth is 20 km).

2) The <u>lithosphere</u> includes the crust and upper part of the <u>mantle</u>, and is made up of a <u>jigsaw</u> of '<u>tectonic plates</u>'. The <u>lithosphere</u> is <u>relatively cold and rigid</u>, and is over 100 km thick in places.

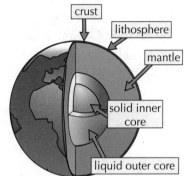

3) The <u>mantle</u> is the <u>solid</u> section between the crust and the core. Near the crust it's <u>very rigid</u>. As you go deeper into the mantle the <u>temperature increases</u> — here it becomes <u>less rigid</u> and can <u>flow very slowly</u> (it behaves like it's semi-liquid).

4) At the centre of the Earth is the <u>core</u>, which we think is made of <u>iron and nickel</u>.

5) The <u>core</u> is just over <u>half</u> the Earth's radius. The <u>inner core</u> is <u>solid</u>, while the <u>outer core</u> is <u>liquid</u>.

6) <u>Radioactive decay</u> creates a lot of the <u>heat</u> inside the Earth. This heat creates <u>convection currents</u> in the mantle, which causes the <u>plates</u> of the lithosphere to <u>move</u>.

The **Earth's Surface** is Made Up of **Tectonic Plates**

1) The crust and the upper part of the mantle are cracked into a number of large pieces called <u>tectonic plates</u>. These plates are a bit like <u>big rafts</u> that 'float' on the mantle.

2) The plates don't stay in one place though. That's because the <u>convection currents</u> in the mantle cause the plates to <u>drift</u>.

3) The map shows the <u>edges</u> of the plates as they are now, and the <u>directions</u> they're moving in (red arrows).

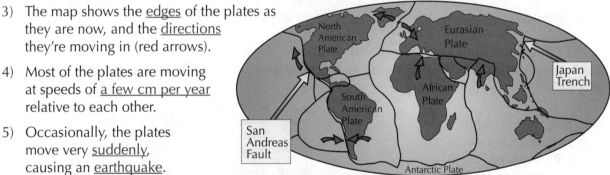

4) Most of the plates are moving at speeds of <u>a few cm per year</u> relative to each other.

5) Occasionally, the plates move very <u>suddenly</u>, causing an <u>earthquake</u>.

6) <u>Volcanoes</u> and <u>earthquakes</u> often occur at the boundaries between two tectonic plates.

Scientists Can't **Predict** Earthquakes and Volcanic Eruptions

1) Tectonic plates can stay more or less put for a while and then <u>suddenly</u> lurch forwards. It's <u>impossible to predict</u> exactly when they'll move.

2) Scientists are trying to find out if there are any <u>clues</u> that an earthquake might happen soon — things like strain in underground rocks. Even with these clues they'll only be able to say an earthquake's <u>likely</u> to happen, not <u>exactly when</u> it'll happen.

3) There are some clues that say a volcanic eruption might happen soon. There's more about this on page 85.

Plate Tectonics

The idea that the Earth's surface is made up of <u>moving plates of rock</u> was proposed by a scientist called <u>Alfred Wegener</u> in the early twentieth century.

Observations About the Earth Hadn't Been Explained

1) For years, fossils of <u>very similar</u> plants and animals had been found on <u>opposite sides</u> of the Atlantic Ocean. Most people thought this was because the continents had been <u>linked</u> by '<u>land bridges</u>', which had <u>sunk</u> or been <u>covered</u> by water as the Earth <u>cooled</u>. But not everyone was convinced, even back then.

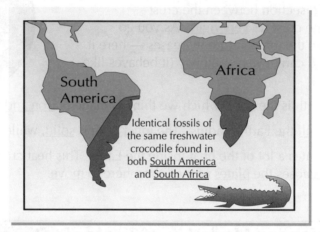

Identical fossils of the same freshwater crocodile found in both <u>South America</u> and <u>South Africa</u>

2) Other things about the Earth puzzled people too — like why the <u>coastlines</u> of <u>Africa</u> and <u>South America</u> fit together and why there are fossils of <u>sea creatures</u> in the <u>Alps</u>.

Explaining These Observations Needed a Leap of Imagination

What was needed was a scientist with a bit of <u>insight</u>... a smidgeon of <u>creativity</u>... a touch of <u>genius</u>...

1) In 1914 <u>Alfred Wegener</u> hypothesised that Africa and South America had previously been <u>one</u> continent which had then <u>split</u>. He started to look for more evidence to <u>back up</u> his hypothesis. He found it...

2) E.g. there were <u>matching layers</u> in the <u>rocks</u> on different continents, and similar <u>earthworms</u> living in <u>both</u> South America and South Africa.

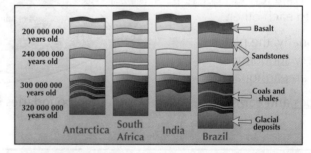

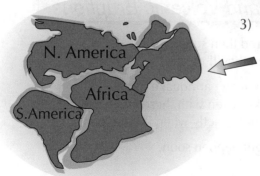

3) Wegener's theory of '<u>continental drift</u>' supposed that about 300 million years ago there had been just one '<u>supercontinent</u>' — which he called Pangaea. According to Wegener, Pangaea broke into smaller chunks, and these chunks (our modern-day <u>continents</u>) are still slowly 'drifting' apart. This idea is the basis behind the modern theory of <u>plate tectonics</u>.

Plate Tectonics

Wegener's theory took a while to <u>catch on</u>, but eventually there was <u>so much evidence</u> that people had to <u>accept</u> what he was saying.

The Theory **Wasn't Accepted at First** — for a **Variety** of Reasons

1) Wegener's theory <u>explained</u> things that <u>couldn't</u> be explained by the 'land bridge' theory (e.g. the formation of <u>mountains</u> — which Wegener said happened as continents <u>smashed into</u> each other). But it was a big change, and the reaction from other scientists was <u>hostile</u>.

2) The main problem was that Wegener's explanation of <u>how</u> the '<u>drifting</u>' happened wasn't convincing (and the movement wasn't <u>detectable</u>). Wegener claimed the continents' movement could be caused by tidal forces and the Earth's rotation — but other geologists showed that this was <u>impossible</u>.

Eventually, the Evidence Became **Overwhelming**

1) In the 1960s, scientists investigated the <u>Mid-Atlantic ridge</u>, which runs the <u>whole length</u> of the Atlantic.

2) They found evidence that <u>magma</u> (molten rock) <u>rises up</u> through the sea floor, <u>solidifies</u> and forms underwater mountains that are <u>roughly symmetrical</u> either side of the ridge. The evidence suggested that the sea floor was <u>spreading</u> — at about 10 cm per year.

3) Even better evidence that the continents are moving apart came from the <u>magnetic orientation</u> of the rocks. As the liquid magma erupts out of the gap, <u>iron particles</u> in the rocks tend to <u>align themselves</u> with the Earth's <u>magnetic field</u> — and as it cools they <u>set</u> in position. Now then... every half million years or so the Earth's magnetic field <u>swaps direction</u> — and the rock on <u>either side</u> of the ridge has <u>bands</u> of <u>alternate magnetic polarity</u>, <u>symmetrical</u> about the ridge.

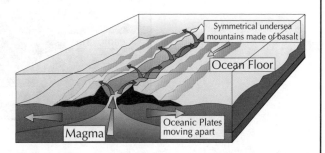

4) This was convincing evidence that new sea floor was being created... and <u>continents</u> were <u>moving apart</u>.

5) All the evidence collected by other scientists <u>supported Wegener's theory</u> — so it was gradually <u>accepted</u>.

Plate Tectonics — it's a smashing theory...

Wegener wasn't right about everything, but his <u>main idea</u> was <u>correct</u>. The scientific community was a bit slow to accept it, but once there was more <u>evidence</u> to support it, they got on board.

84

Volcanic Eruptions

The theory of plate tectonics not only explains why the continents move, it also makes sense of natural hazards such as volcanoes and earthquakes.

Volcanoes are Formed by *Molten Rock*

1) Volcanoes occur when molten rock (magma) from the mantle emerges through the Earth's crust.

2) Magma rises up (through the crust) and 'boils over' where it can — sometimes quite violently if the pressure is released suddenly.

> When the molten rock is below the surface of the Earth it's called magma — but when it erupts from a volcano it's called lava.

Oceanic and *Continental Crust Colliding* Causes Volcanoes

1) The crust at the ocean floor is denser than the crust below the continents.

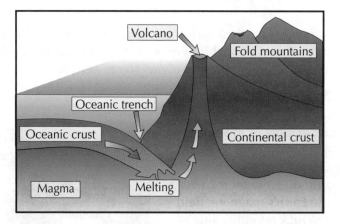

2) When two tectonic plates collide, a dense oceanic plate will be forced underneath a less dense continental plate. This is called subduction.

3) Oceanic crust also tends to be cooler at the edges of a tectonic plate — so the edges sink easily, pulling the oceanic plate down.

4) As the oceanic crust is forced down it melts and starts to rise. If this molten rock finds its way to the surface, volcanoes form.

Volcanic Eruptions

Volcanic Activity Forms *Igneous Rock*

1) Igneous rock is made when any sort of molten rock cools down and solidifies. Lots of rocks on the surface of the Earth were formed this way.

2) The type of igneous rock (and the behaviour of the volcano) depends on how quickly the magma cools and the composition of the magma.

3) Some volcanoes produce magma that forms iron-rich basalt. The lava from the eruption is runny, and the eruption is fairly safe. (As safe as you can be with molten rock at 1200 °C, I suppose.)

4) But if the magma is silica-rich rhyolite, the eruption is explosive. It produces thick lava which can be violently blown out of the top of the volcano.

Geologists *Try* to *Predict Volcanic Eruptions*

1) Geologists study volcanoes to try to find out if there are signs that a volcanic eruption might happen soon — things like magma movement below the ground near to a volcano. This causes mini-earthquakes near the volcano.

2) Being able to spot these kinds of clues means that scientists can predict eruptions with much greater accuracy than they could in the past.

3) It's tricky though — volcanoes are very unpredictable. For example, sometimes molten rock cools down instead of erupting, so mini-earthquakes can be a false alarm.

4) Most likely, scientists will only be able to say that an eruption's more likely than normal — not that it's certain. But even just knowing that can save lives.

Volcanoes — they're unpredictable

Volcanoes can erupt with huge force, so it might seem odd to choose to live near one. But there are benefits — volcanic ash creates very fertile soil that's great for farming. It'd be much safer if eruptions could be predicted accurately — they're not perfect yet, but predictions are getting better all the time.

The Three Different Types of Rock

Scientists classify rocks according to how they're formed. The three different types are: <u>sedimentary</u>, <u>metamorphic</u> and <u>igneous</u>. Sedimentary rocks are generally pretty soft, while igneous rocks are hard.

*There are **Three Steps** in the Formation of **Sedimentary Rock***

1) <u>Sedimentary rocks</u> are formed from <u>layers of sediment</u> laid down in <u>lakes</u> or <u>seas</u>.

2) Over <u>millions of years</u> the layers get <u>buried</u> under more layers and the <u>weight</u> pressing down <u>squeezes out</u> the water.

3) Fluids flowing through the pores deposit natural mineral <u>cement</u>.

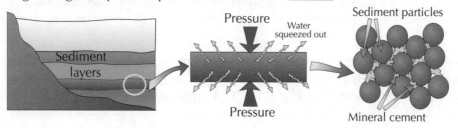

*Limestone is a Sedimentary Rock Formed from **Seashells***

1) Limestone is mostly formed from <u>seashells</u>. It's mostly <u>calcium carbonate</u> and <u>grey/white</u> in colour. The original <u>shells</u> are mostly <u>crushed</u>, but there can still be quite a few <u>fossilised shells</u> remaining.

2) When limestone is heated it <u>thermally decomposes</u> to make <u>calcium oxide</u> and <u>carbon dioxide</u>:

calcium carbonate → calcium oxide + carbon dioxide
$$CaCO_{3(s)} \quad \rightarrow \quad CaO_{(s)} \quad + \quad CO_{2(g)}$$

Thermal decomposition is when one substance chemically changes into at least two new substances when it's heated.

*Metamorphic rocks are Formed from **Other Rocks***

1) <u>Metamorphic rocks</u> are formed by the action of <u>heat and pressure</u> on <u>sedimentary</u> (or even <u>igneous</u>) <u>rocks</u> over <u>long periods</u> of time.

2) The <u>mineral structure</u> and <u>texture</u> may be different, but the chemical composition is often the same.

3) So long as the rocks don't actually <u>melt</u> they're classed as <u>metamorphic</u>. If they <u>melt</u> and turn to <u>magma</u>, they're <u>gone</u> (though they may eventually resurface as igneous rocks).

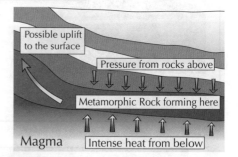

*Marble is a Metamorphic Rock Formed from **Limestone***

1) Marble is another form of <u>calcium carbonate</u>.

2) Very high temperatures and pressures <u>break down</u> the limestone and it reforms as <u>small crystals</u>.

3) This gives marble a <u>more even texture</u> and makes it <u>much harder</u>.

*Igneous Rocks are Formed from **Fresh Magma***

1) <u>Igneous rocks</u> are formed when <u>magma</u> cools.

2) They contain various <u>different minerals</u> in <u>randomly arranged</u> interlocking <u>crystals</u> — this makes them very <u>hard</u>.

3) Granite is a <u>very hard</u> igneous rock (even harder than marble). It's ideal for <u>steps</u> and <u>buildings</u>.

The Evolution of the Atmosphere

For 200 million years or so, the atmosphere has been about how it is now: approximately 78% nitrogen, 21% oxygen, 1% argon and small amounts of other gases, mainly carbon dioxide, noble gases and water vapour. But it wasn't always like this. Here's how the past 4.5 billion years may have gone:

Phase 1 — *Volcanoes* Gave Out *Gases*

1) The Earth's surface was originally <u>molten</u> for many millions of years. It was so hot that any atmosphere just '<u>boiled away</u>' into space.

2) Eventually things cooled down a bit and a <u>thin crust</u> formed, but <u>volcanoes</u> kept erupting.

3) The volcanoes gave out lots of gas. We think this was how the oceans and atmosphere were formed.

4) The early atmosphere was probably <u>mostly CO_2</u>, with virtually <u>no oxygen</u>. There may also have been <u>water vapour</u>, and small amounts of <u>methane</u> and <u>ammonia</u>. This is quite like the atmospheres of Mars and Venus today.

5) The <u>oceans</u> formed when the water vapour <u>condensed</u>.

<u>Holiday report</u>: Not a nice place to be. Take strong walking boots and a good coat.

Phase 2 — *Green plants* Evolved and Produced *Oxygen*

<u>Holiday report</u>: A bit slimy underfoot. Take wellies and a lot of suncream.

1) <u>Green plants</u> and <u>algae</u> evolved over most of the Earth. They were quite happy in the <u>CO_2 atmosphere</u>.

2) A lot of the early CO_2 <u>dissolved</u> into the oceans. The <u>green plants</u> and <u>algae</u> also absorbed some of the <u>CO_2</u> and <u>produced O_2</u> by <u>photosynthesis</u>.

3) Plants and algae died and were buried under layers of sediment, along with the skeletons and shells of marine organisms that had slowly evolved. The <u>carbon</u> and <u>hydrocarbons</u> inside them became 'locked up' in <u>sedimentary rocks</u> as <u>insoluble carbonates</u> (e.g. limestone) and <u>fossil fuels</u>.

4) When we <u>burn</u> fossil fuels today, this 'locked-up' carbon is released and the concentration of CO_2 in the atmosphere rises.

Phase 3 — *Ozone layer* Allows Evolution of *Complex Animals*

1) The build-up of <u>oxygen</u> in the atmosphere <u>killed off</u> some early organisms that couldn't tolerate it, but allowed other, more complex organisms to evolve and flourish.

2) The oxygen also created the <u>ozone layer</u> (O_3) which <u>blocked</u> harmful rays from the Sun and <u>enabled</u> even <u>more complex</u> organisms to evolve — us, eventually.

3) There is virtually <u>no CO_2</u> left now.

<u>Holiday report</u>: A nice place to be. Visit before the crowds ruin it.

Life, Resources and Atmospheric Change

Life on Earth began <u>billions of years</u> ago, but there's no way of knowing for definite how it all started.

*Primordial Soup is Just One **Theory** of How **Life** Was **Formed***

1) The primordial soup theory states that billions of years ago, the Earth's <u>atmosphere</u> was rich in <u>nitrogen</u>, <u>hydrogen</u>, <u>ammonia</u> and <u>methane</u>.

2) <u>Lightning</u> struck, causing a chemical reaction between the gases, resulting in the formation of <u>amino acids</u>.

3) The amino acids collected in a '<u>primordial soup</u>' — a body of water out of which life gradually crawled.

4) The amino acids gradually combined to produce <u>organic matter</u> which eventually evolved into simple <u>living organisms</u>.

5) In the 1950s, <u>Miller and Urey</u> carried out an experiment to prove this theory. They sealed the gases in their apparatus, heated them and applied an electrical charge for a week.

6) They found that <u>amino acids were made</u>, but not as many as there are on Earth. This suggests the theory could be along the <u>right lines</u>, but isn't quite right.

*The **Earth** Has All the **Resources** Humans Need*

The Earth's crust, oceans and atmosphere are the <u>ultimate source</u> of minerals and resources — we can get everything we need from them. For example, we can <u>fractionally distil air</u> to get a variety of products (e.g. nitrogen and oxygen) for use in <u>industry</u>:

1) Air is <u>filtered</u> to remove dust.

2) It's then <u>cooled</u> to around <u>-200 °C</u> and becomes a liquid.

3) During cooling <u>water vapour</u> condenses and is removed.

4) <u>Carbon dioxide</u> freezes and is removed.

5) The liquefied air then enters the fractionating column and is <u>heated</u> slowly.

6) The remaining gases are separated by <u>fractional distillation</u>. Oxygen and argon come out together so <u>another</u> column is used to separate them.

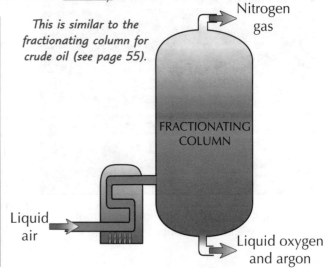

This is similar to the fractionating column for crude oil (see page 55).

Nitrogen gas

FRACTIONATING COLUMN

Liquid air

Liquid oxygen and argon

*Carbon Dioxide Level Affects the **Climate** and the **Oceans***

<u>Burning fossil fuels</u> releases CO_2 — and as the world's become more industrialised, more fossil fuels have been burnt in power stations and in car engines. This CO_2 is thought to be altering our planet...

1) An increase in carbon dioxide is causing <u>global warming</u> — a type of <u>climate change</u> (see p.93).

2) The oceans are a <u>natural store</u> of CO_2 — they absorb it from the atmosphere. However the extra CO_2 we're releasing is making them too <u>acidic</u>. This is bad news for <u>coral</u> and <u>shellfish</u>, and also means that in the future they won't be able to absorb any more carbon dioxide.

Warm-Up and Exam Questions

If you still think the Earth is flat, you may want to re-read the last few pages. If you think you know otherwise, and also know more interesting facts about the Earth's structure, test yourself with these...

Warm-Up Questions

1) State one geological feature often seen at the boundary of two tectonic plates.
2) What was 'Pangaea'?
3) Explain why an oceanic plate will be pushed below a continental plate when they collide.
4) Name three different types of rock.
5) Where did the gases that made up the early atmosphere come from?
6) What is the name of the process used to separate the gases in air?

Exam Questions

1   The following diagram shows the internal structure of the Earth.

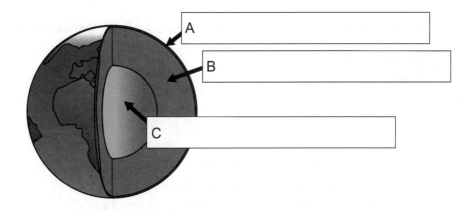

    A

    B

    C

(a)   Label the diagram.

(3 marks)

(b)   The part labelled A is cracked into many pieces.
    (i)   What are these pieces called?

(1 mark)

    (ii)   Explain the process that causes these pieces to move.

(3 marks)

2   Alfred Wegener came up with the theory of continental drift.
(a)   Describe two pieces of evidence to support his theory.

(2 marks)

(b)   Give one reason why it wasn't accepted at the time.

(1 mark)

Exam Questions

3   Metamorphic rocks, such as marble, are formed from other types of rock.
   (a) Describe how metamorphic rocks are formed from sedimentary rocks.

(2 marks)

   (b) (i)   What sedimentary rock is marble formed from?

(1 mark)

       (ii) Give two differences between the physical properties of marble
             and those of the rock it is formed from.

(2 marks)

   (c) Describe how sedimentary rocks are formed.

(3 marks)

4   Igneous rocks are formed from magma.
   (a) Explain the difference between magma and lava.

(2 marks)

   (b) Describe how the structure of igneous rocks means that they are very hard.

(1 mark)

   (c) Give one example of an igneous rock.

(1 mark)

5   Match the words for **A**, **B**, **C** and **D** with the numbers **1 - 4** in the sentences below.

       A   photosynthesis
       B   oxygen
       C   carbon
       D   carbon dioxide

   Once green plants had evolved, they thrived in an atmosphere rich in ...**1**.... .
   These plants produced ...**2**... by the process of ...**3**.... .
   ...**4**... from dead plants eventually became 'locked up' in fossil fuels.

(4 marks)

6   The composition of the Earth's atmosphere was vital to the development of life on Earth.
   (a) Explain the 'primordial soup' theory of how life on Earth was formed.

(4 marks)

   (b) Describe an experiment that was conducted to test the 'primordial soup'
        theory of how life on Earth was formed.

(3 marks)

   (c) What percentage of the present day atmosphere is made up of oxygen?

(1 mark)

7   Air can be separated out into its different parts.
   Describe the process used to separate out the gases in the air.

(6 marks)

Air Pollution — Carbon and Sulfur

90% of crude oil is used as fuel. It's burnt to release the energy stored inside it.

Burning Fossil Fuels Releases Gases and Particles

1) Power stations burn huge amounts of fossil fuels to make electricity.
   Cars are also a major culprit in burning fossil fuels.

2) Most fuels, such as crude oil and coal, contain carbon and hydrogen.
   During combustion, the carbon and hydrogen are oxidised so that
   carbon dioxide and water vapour are released into the atmosphere. Energy (heat) is also produced.
   E.g.:

   hydrocarbon + oxygen → carbon dioxide + water vapour

3) When there's plenty of oxygen, all the fuel burns — this is called complete combustion.

4) If there's not enough oxygen, some of the fuel doesn't burn — this is called partial combustion.
   Under these conditions, solid particles (called particulates) of soot (carbon) and unburnt fuel
   are released. Carbon monoxide (a poisonous gas) is also released.

   *Partial combustion
   is also known as
   incomplete combustion.*

Sulfur Dioxide Causes Acid Rain

1) Sulfur dioxide is one of the gases that causes acid rain.

2) When the sulfur dioxide mixes with clouds it forms dilute sulfuric acid. This then falls as acid rain.

3) In the same way, oxides of nitrogen cause acid rain by forming dilute nitric acid in clouds.

4) Acid rain causes lakes to become acidic and many plants and animals die as a result.

5) Acid rain kills trees and damages limestone buildings and ruins stone statues. It's shocking.

6) Links between acid rain and human health problems have been suggested.

7) The benefits of electricity and travel have to be balanced against the environmental impacts.
   Governments have recognised the importance of this and international agreements have
   been put in place to reduce emissions of air pollutants such as sulfur dioxide.

You Can Reduce Acid Rain by Reducing Sulfur Emissions

1) Most of the sulfur can be removed from fuels before they're burnt, but it costs more to do it.

2) Also, removing sulfur from fuels takes more energy. This usually comes from burning more fuel,
   which releases more of the greenhouse gas carbon dioxide.

3) However, petrol and diesel are starting to be replaced by low-sulfur versions.

4) Power stations now have Acid Gas Scrubbers to take the harmful gases out before
   they release their fumes into the atmosphere.

5) The other way of reducing acid rain is simply to reduce our usage of fossil fuels.

Air Pollution — Nitrogen

Now for a new type of pollution. One that, surprisingly, is made from the <u>nitrogen</u> in the <u>air</u> itself.

Nitrogen Pollution Involves Nitrogen from the Air

Nitrogen pollution <u>doesn't</u> actually come from the <u>fuel itself</u>
— it's formed from nitrogen in the <u>air</u> when the <u>fuel is burnt</u>.

1) Fossil fuels burn at such <u>high temperatures</u> that nearby <u>atoms</u> in the air <u>react</u> with each other.

2) <u>Nitrogen</u> in the air reacts with the <u>oxygen</u> in the air to produce small amounts of compounds known as <u>nitrogen oxides</u> — <u>nitrogen monoxide</u> and <u>nitrogen dioxide</u>.

3) This happens in <u>car engines</u>.

4) Nitrogen oxides are <u>pollutants</u>, and are usually spewed straight out into the <u>atmosphere</u>.

*Nitrogen Oxides Are **Nitrogen Monoxide** and **Nitrogen Dioxide***

1) <u>Nitrogen monoxide</u> has the formula NO — it is made of <u>one nitrogen</u> and <u>one oxygen</u> atom.

2) <u>Nitrogen dioxide</u> has the formula NO_2 — it is made of <u>two oxygen</u> atoms joined to <u>one nitrogen</u> atom.

3) <u>Nitrogen oxides</u> (NO and NO_2) can be jointly referred to as NO_x.

Nitrogen monoxide (NO)

Nitrogen dioxide (NO_2)

Here's how nitrogen oxides are <u>formed</u>:

Nitrogen Monoxide

<u>Nitrogen monoxide</u> forms when <u>nitrogen</u> and <u>oxygen</u> in the air are exposed to a very <u>high temperature</u>.
This happens when fuels are burnt in places like car <u>engines</u>.

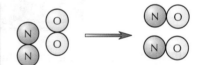

These are both oxidation reactions.

Nitrogen Dioxide

Once the nitrogen monoxide is in the air, it will go on to react with more <u>oxygen</u> in the air to produce <u>nitrogen dioxide</u>.

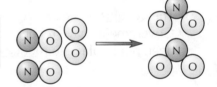

As pollutants, nitrogen oxides are very similar to <u>sulfur dioxide</u>.
When they're formed they usually end up in the atmosphere — which is where they stay until they <u>react</u> with <u>moisture</u> in clouds. This produces a <u>dilute nitric acid</u> which eventually falls to the Earth as <u>acid rain</u>.

A pollution solution would be nice...

Nitrogen pollution is another type of pollution to worry about. It would be a lot easier if we all went back to the old days of farming a bit of land in a self-sufficient manner. No more electricity, no more industry. Although there'd be no more TVs, DVD players, X-ray machines, computers. Hmm...

Environmental Problems

More doom and gloom on this page I'm afraid... but we all need to be aware of it.

Increasing Carbon Dioxide Causes Climate Change

1) The level of <u>carbon dioxide</u> in the atmosphere is <u>increasing</u> —
   because of the large amounts of <u>fossil fuels</u> humans burn.

2) There's a <u>scientific consensus</u> that this extra carbon dioxide has caused
   the average <u>temperature</u> of the Earth to <u>increase</u> — <u>global warming</u>.

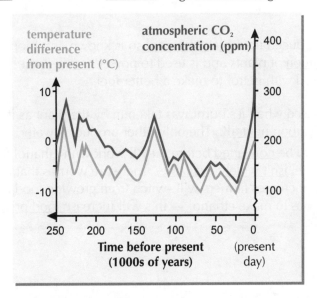

3) Global warming is a type of <u>climate change</u> and causes other types of climate change, e.g.
   changing rainfall patterns. It could also cause severe <u>flooding</u> due to the polar ice caps melting.

Particles Cause Global Dimming

1) In the last few years, some scientists have been measuring how much <u>sunlight</u> is reaching
   the surface of the Earth and comparing it to records from the last 50 years.

2) They have been amazed to find that in some areas nearly <u>25% less sunlight</u> has been
   reaching the surface compared to 50 years ago. They have called this <u>global dimming</u>.

3) They think that it is caused by <u>particles</u> of soot and ash that are
   produced when <u>fossil fuels</u> are burnt. These particles <u>reflect</u> sunlight
   back into space, or they can help to produce more <u>clouds</u> that reflect
   the sunlight back into space.

4) There are many scientists who <u>don't believe</u> the change is
   real and blame it on <u>inaccurate</u> recording equipment.

*London by day —
if global dimming
gets really bad.*

Global dimming — romantic lighting all day...

On a cold winter's day, I often think that a bit of global warming would be nice — but it seems that
it does <u>mess things up</u>. There have been lots of other times in the past when the climate has <u>changed</u>
— volcanic eruptions, changes in the Earth's orbit and the movements of tectonic plates (see page 81)
have all had effects. But this is the <u>first time</u> we've had the technology and knowledge to investigate it.

Environmental Problems

As the demand on resources increases it's important to develop new, <u>alternative fuels</u>.

Alternative fuels are Being Developed

Some <u>alternative fuels</u> have already been developed, and there are others in the pipeline (so to speak). Many of them are <u>renewable</u> fuels so, unlike fossil fuels, they won't run out. However, none of them are perfect — they all have <u>pros and cons</u>. For example:

Ethanol

ETHANOL can be produced from <u>plant material</u> so is known as a <u>biofuel</u>. It's made by <u>fermentation</u> of plants and is used to power <u>cars</u> in some places. It's often mixed with petrol to make a better fuel.

<u>PROS</u>: The CO_2 released when it's burnt was taken in by the plant as it grew, so it's '<u>carbon neutral</u>'. The only other product is <u>water</u>.

<u>CONS</u>: Engines need to be <u>converted</u> before they'll work with ethanol fuels. And ethanol fuel <u>isn't widely available</u>. There are worries that as demand for it increases farmers will switch from growing food crops to growing crops to make ethanol — this will <u>increase food prices</u>.

Biodiesel

BIODIESEL is another type of <u>biofuel</u>. It can be produced from <u>vegetable oils</u> such as rapeseed oil and soybean oil. Biodiesel can be mixed with ordinary diesel fuel and used to run a <u>diesel engine</u>.

<u>PROS</u>: Biodiesel is '<u>carbon neutral</u>'. <u>Engines don't</u> need to be <u>converted</u>. It produces much <u>less sulfur dioxide</u> and '<u>particulates</u>' than ordinary diesel or petrol.

<u>CONS</u>: We <u>can't make enough</u> to completely replace diesel. It's <u>expensive</u> to make. It could <u>increase food prices</u> like using more ethanol could (see above).

Hydrogen gas

HYDROGEN GAS can also be used to power vehicles. You get the hydrogen from the <u>electrolysis of water</u> — there's plenty of water about but it takes <u>electrical energy</u> to split it up. This energy can come from a <u>renewable</u> source, e.g. solar.

<u>PROS</u>: Hydrogen combines with oxygen in the air to form <u>just water</u> — so it's <u>very clean</u>.

<u>CONS</u>: You need a <u>special, expensive engine</u> and hydrogen <u>isn't widely available</u>. You still need to use <u>energy</u> from <u>another source</u> to make it. Also, hydrogen's hard to <u>store</u>.

We should probably be using alternative fuels already...

You may have heard about these fuels in the news. Some of the big car manufacturers are spending loads of money trying to make hydrogen-powered cars a reality. These are the fuels of the future...

Warm-Up and Exam Questions

Question time again. Oh joy. Boring I know, but a great way of testing what you do and don't know.

Warm-Up Questions

1) What are Acid Gas Scrubbers used for?
2) What is the formula for nitrogen monoxide?
3) Describe how nitrogen dioxide is produced when fossil fuels are burnt.
4) What is a renewable fuel?
5) Name three alternative fuels.

Exam Questions

1   When fossil fuels like petrol are burnt, they produce carbon dioxide, sulfur dioxide and particulate matter.
   (a)   Name the atmospheric environmental problem caused by increased levels of carbon dioxide.
(1 mark)
   (b)   (i)   Describe how acid rain is formed.
(3 marks)
         (ii)   Give **one** effect of acid rain.
(1 mark)
   (c)   (i)   What is global dimming?
(1 mark)
         (ii)   What causes global dimming?
(1 mark)
   (d)   One advantage of burning ethanol is that it produces no sulfur.
         Give **two** disadvantages of using ethanol as a fuel compared to petrol.
(2 marks)

2   Currently, fossil fuels provide about 60-70% of the world's electricity.

   Since fossils fuels will eventually run out, it is important to find alternative energy sources for the future. An example of an alternative fuel is biodiesel. It can be used in diesel engines.
   (a)   What is biodiesel made from?
(1 mark)
   (b)   (i)   Explain why biodiesel doesn't add to global warming.
(2 marks)
         (ii)   Give **two** other advantages of biodiesel.
(2 marks)
   (c)   Give **two** disadvantages of biodiesel.
(2 marks)

3   Scientists are developing 'fuels for the future'. One example could be hydrogen.
   (a)   Describe how hydrogen is produced on a large scale.
(1 mark)
   (b)   Give **one** advantage of using hydrogen as a fuel.
(1 mark)
   (c)   Give **one** disadvantage of using hydrogen as a fuel in a car compared with using petrol or diesel.
(1 mark)

Revision Summary for Section 5

The only way that you can tell if you've learned this section properly is to test yourself. Try these questions, and if there's something you don't know, it means you need to go back and learn it. Don't miss any questions out — that'd be cheating.

1) Briefly describe the inner structure of the Earth.

2) A geologist places a very heavy marker on the seabed in the middle of the Atlantic ocean. She records the marker's position over a period of four years. The geologist finds that the marker moves in a straight-line away from its original position. Her measurements are shown in the graph below.

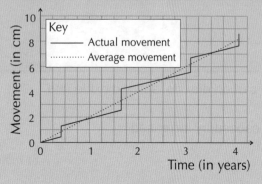

a) Explain the process that has caused the marker to move.

b)*What is the marker's average movement each year?

c)*On average, how many years will it take for the marker to move 7 cm?

3) Why can't scientists accurately predict volcanoes and earthquakes?

4) Briefly describe Wegener's theory of continental drift.

5) What is meant by 'subduction'?

6) Sketch a labelled diagram showing how a volcano forms.

7) How is an eruption of silica-rich rhyolitic lava different from an eruption of iron-rich basaltic lava?

8) Draw a diagram to show how metamorphic rocks form.

9) Give an example of a metamorphic rock and say what the material it formed from is.

10) Which material is hardest, granite, limestone or marble?

11) Name the three main gases that make up the Earth's atmosphere today.

12) Explain why today's atmosphere is different from the Earth's early atmosphere.

13) What is meant by 'primordial soup'?

14) Why do we fractionally distil air?

15) The burning of fossils fuels is causing a rise in the level of carbon dioxide in the atmosphere. How is this affecting the oceans and the climate?

16) Describe briefly how the pollutant sulfur dioxide is produced.

17) What effects does acid rain have on the environment?

18) List three ways of reducing acid rain.

19) What effect do nitrogen oxides have on the environment?

20) List three alternative ways of powering cars. What are the pros and cons of each?

* Answers on page 254

History of the Periodic Table

We haven't always known as much about Chemistry as we do now. Early chemists looked to try and understand <u>patterns</u> in the elements' properties to get a bit of understanding.

Döbereiner *Tried to Organise Elements into* Triads

1) Back in the 1800s the only thing they could measure was <u>relative atomic mass</u>, and so the <u>known</u> elements were arranged <u>in order of atomic mass</u>.

2) In 1828 a guy called <u>Döbereiner</u> started to put this list of elements into groups based on their <u>chemical properties</u>. He put the elements into groups of <u>three</u>, which he called <u>triads</u>. E.g. Cl, Br and I were one triad, and Li, Na and K were another.

3) The <u>middle element</u> of each triad had a relative atomic mass that was the <u>average</u> of the other two.

Element	Relative atomic mass
Lithium	7
Sodium	23
Potassium	39

(7 + 39) ÷ 2 = 23

Newlands' Law of Octaves *Was the First Good Effort*

A chap called <u>Newlands</u> noticed that when you arranged the elements in order of relative atomic mass, every <u>eighth</u> element had similar properties, and so he listed some of the known elements in rows of seven:

H	Li	Be	B	C	N	O
F	Na	Mg	Al	Si	P	S
Cl	K	Ca	Cr	Ti	Mn	Fe

These sets of eight were called <u>Newlands' Octaves</u>. Unfortunately the pattern <u>broke down</u> on the <u>third row</u>, with <u>transition metals</u> like titanium (Ti) and iron (Fe) messing it up.

It was because he left <u>no gaps</u> that his work was <u>ignored</u>. But he was getting <u>pretty close</u>.

Newlands presented his ideas to the Chemical Society in 1865. But his work was criticised because:

1) His groups contained elements that didn't have <u>similar properties</u>, e.g. <u>carbon</u> and <u>titanium</u>.

2) He <u>mixed up metals and non-metals</u> e.g. <u>oxygen</u> and <u>iron</u>.

3) He <u>didn't leave any gaps</u> for elements that hadn't been discovered yet.

History of the Periodic Table

Newlands wasn't the only one who had ideas about classifying elements.

Dmitri Mendeleev Left Gaps and Predicted New Elements

1) In 1869, Dmitri Mendeleev in Russia, armed with about 50 known elements, arranged them into his Table of Elements — with various gaps as shown.

<u>Mendeleev's Table of the Elements</u>

```
H
Li  Be                                          B   C   N   O   F
Na  Mg                                          Al  Si  P   S   Cl
K   Ca  *  Ti  V  Cr  Mn  Fe  Co  Ni  Cu  Zn  *   *   As  Se  Br
Rb  Sr  Y  Zr  Nb Mo  *   Ru  Rh  Pd  Ag  Cd  In  Sn  Sb  Te  I
Cs  Ba  *  *   Ta W   *   Os  Ir  Pt  Au  Hg  Tl  Pb  Bi
```

2) Mendeleev put the elements in order of atomic mass (like Newlands did).

3) But Mendeleev found he had to leave gaps in order to keep elements with similar properties in the same vertical groups — and he was prepared to leave some very big gaps in the first two rows before the transition metals come in on the third row.

4) The gaps were the really clever bit because they predicted the properties of so far undiscovered elements.

5) When they were found and they fitted the pattern it helped confirm Mendeleev's ideas. For example, Mendeleev made really good predictions about the chemical and physical properties of an element he called ekasilicon, which we know today as germanium.

Not All Scientists Thought the Periodic Table was Important

1) When the periodic table was first released, many scientists thought it was just a bit of fun. At that time, there wasn't all that much evidence to suggest that the elements really did fit together in that way — ideas don't get the scientific stamp of approval without evidence.

2) After Mendeleev released his work, newly discovered elements fitted into the gaps he left. This was convincing evidence in favour of the periodic table.

3) Once there was more evidence, many more scientists realised that the periodic table could be a useful tool for predicting properties of elements. It really worked.

4) In the late 19th century, scientists discovered protons, neutrons and electrons. The periodic table matches up very well to what's been discovered about the structure of the atom. Scientists now accept that it's a very important and useful summary of the structure of atoms.

Elementary my dear Mendeleev

Mendeleev tried to classify the 50 elements that were known at the time — and although his attempt wasn't perfect, he got on pretty well. It was an improvement on Newlands' attempt anyway.

The Modern Periodic Table

Chemists were getting pretty close to producing something useful.
The big breakthrough came when the structure of the atom was understood a bit better.

The *Modern Periodic Table* is Based on *Electronic Structure*

When electrons, protons and neutrons were discovered, the periodic table was arranged
in order of atomic number. All elements were put into groups.

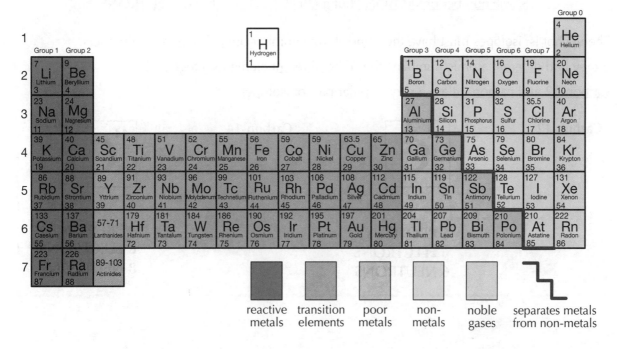

1) The elements in the periodic table can be seen as being arranged by their electronic structure.
   Using the electron arrangement, you can predict the element's chemical properties.

2) Electrons in an atom are set out in shells which each correspond to an energy level.

3) Apart from the transition metals, elements in the same group have the same number of electrons
   in their highest occupied energy level (outer shell).

4) The group number is equal to the number of electrons in the highest occupied energy level
   — e.g. Group 6 all have 6 electrons in the highest energy level.

5) The positive charge of the nucleus attracts electrons and holds them in place.
   The further from the nucleus the electron is, the less the attraction.

6) The attraction of the nucleus is even less when there are a lot of inner electrons.
   Inner electrons "get in the way" of the nuclear charge, reducing the attraction.
   This effect is known as shielding.

7) The combination of increased distance and increased shielding means that an electron
   in a higher energy level is more easily lost because there's less attraction from the nucleus
   holding it in place. That's why Group 1 metals get more reactive as you go down the group.

8) Increased distance and shielding also means that a higher energy level is less likely
   to gain an electron — there's less attraction from the nucleus pulling electrons into
   the atom. That's why Group 7 elements get less reactive going down the group.

Ahh, finally, the nice colourful periodic table in full...

This is a good example of how science often progresses — even now. A scientist has a basically good
(though incomplete) idea. Other scientists laugh and mock and generally deride. Eventually, the idea
is modified a bit to take account of the available evidence, and into the textbooks it goes.

Isotopes and Relative Atomic Mass

Some elements have more than one isotope.

Isotopes Are The Same Except For An Extra Neutron Or Two

Here's the definition of an isotope:

> Isotopes are: different atomic forms of the same element, which have the SAME number of PROTONS but a DIFFERENT number of NEUTRONS.

1) The upshot is: isotopes must have the same atomic number but different mass numbers.

2) If they had different atomic numbers, they'd be different elements altogether.

3) Carbon-12 and carbon-14 are a very popular pair of isotopes.

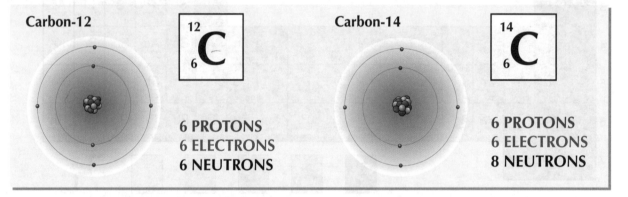

Carbon-12

$^{12}_{\ 6}C$

6 PROTONS
6 ELECTRONS
6 NEUTRONS

Carbon-14

$^{14}_{\ 6}C$

6 PROTONS
6 ELECTRONS
8 NEUTRONS

Relative Atomic Mass Takes Isotopes into Account

1) Relative atomic mass (A_r) uses the average mass of all the isotopes of an element. It has to allow for the relative mass of each isotope and its relative abundance.

2) Relative abundance just means how much there is of each isotope compared to the total amount of the element in the world. This can be a ratio, a fraction or a percentage.

EXAMPLE: Work out the relative atomic mass of chlorine.

element	relative mass of isotope	relative abundance
chlorine	35	3
	37	1

ANSWER:
This means that there are two isotopes of chlorine. One has a relative mass of 35 (^{35}Cl) and the other has a relative mass of 37 (^{37}Cl).

The relative abundances of the two isotopes show that there are three atoms of ^{35}Cl for every one atom of ^{37}Cl.

- First, multiply the mass of each isotope by its relative abundance.
- Add those numbers together.
- Divide by the sum of the relative abundances.

$$A_r = \frac{(35 \times 3) \ + \ (37 \times 1)}{3 \ + \ 1} = \underline{35.5}$$

3) Relative atomic masses don't usually come out as whole numbers or easy decimals, but they're often rounded to the nearest 0.5 in periodic tables (see page 99).

Warm-Up and Exam Questions

Try your hand at these questions — it's the best way to test whether you know your stuff.
If there's anything you can't answer, be sure to go back through the section and look it up.

Warm-Up Questions

1) What were Döbereiner's triads?
2) How did Mendeleev's Table of Elements improve on Newlands' attempt?
3) In the modern periodic table, name one thing that you can tell about an element from its group number.
4) What is the name given to atoms of the same element with different mass numbers?

Exam Questions

1   Carbon has several isotopes, for example carbon-12 and carbon-13.
    Details about the carbon-13 isotope are shown below:

$$^{13}_{6}\text{C}$$

(a) Explain what an isotope is.

(3 marks)

(b) Draw a diagram to represent the carbon-13 atom. Label the number of protons and neutrons in the nucleus and show the electron arrangement.

(3 marks)

2   Boron has two main isotopes, $^{11}_{5}\text{B}$ and $^{10}_{5}\text{B}$. Its A_r value is 10.8.

(a) What does A_r stand for?

(1 mark)

(b) What is the difference between the two boron isotopes?

(1 mark)

(c) Which isotope is the most abundant? Explain your reasoning.

(2 marks)

3   Which of the following statements about Mendeleev's periodic table is **not** true?

   A   Elements with similar chemical properties were placed in vertical groups.
   B   Gaps were left which helped in predicting the properties of undiscovered elements.
   C   The elements were arranged in order of atomic mass.
   D   Mendeleev's periodic table contained over 100 elements.

(1 mark)

4   Newlands and Mendeleev both came up with a system for classifying elements.

(a) Describe Newlands' system for classifying elements.

(1 mark)

(b) Explain why the work of Mendeleev was taken more seriously than that of Newlands.

(4 marks)

Ionic Bonding

Ionic bonding is one of the ways atoms can form compounds.

Ionic Bonding — Transferring Electrons

In ionic bonding, atoms lose or gain electrons to form charged particles (called ions) which are then strongly attracted to one another (because of the attraction of opposite charges, + and –).

A Shell with Just One Electron is Well Keen to Get Rid...

1) All the atoms over at the left-hand side of the periodic table, e.g. sodium, potassium, calcium etc. have just one or two electrons in their outer shell (highest energy level).

2) And they're pretty keen to get shot of them, because then they'll only have full shells left, which is how they like it. (They try to have the same electronic structure as a noble gas.)

3) So given half a chance they do get rid, and that leaves the atom as an ion instead.

4) Now ions aren't the kind of things that sit around quietly watching the world go by. They tend to leap at the first passing ion with an opposite charge and stick to it like glue.

A Nearly Full Shell is Well Keen to Get That Extra Electron...

1) On the other side of the periodic table, the elements in Group 6 and Group 7, such as oxygen and chlorine, have outer shells which are nearly full.

2) They're obviously pretty keen to gain that extra one or two electrons to fill the shell up.

3) When they do, of course, they become ions and before you know it, pop, they've latched onto the atom (ion) that gave up the electron a moment earlier.

The reaction of sodium and chlorine is a classic case:

The sodium atom gives up its outer electron and becomes an Na⁺ ion.

The chlorine atom has picked up the spare electron and becomes a Cl⁻ ion.

POP!

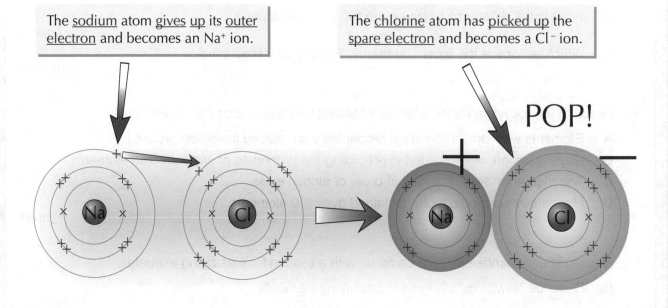

*A diagram like this that shows the electron arrangements
of two atoms is sometimes called a 'dot and cross' diagram.*

Ionic Bonding

Ionic bonds produce <u>giant ionic structures</u>.

Ionic Compounds Have a Regular Lattice Structure

1) <u>Ionic compounds</u> always have <u>giant ionic lattices</u>.

2) The ions form a closely packed <u>regular lattice</u> arrangement.

3) There are very strong <u>electrostatic forces of attraction</u>
   between <u>oppositely charged</u> ions, in <u>all directions</u>.

4) A single crystal of <u>sodium chloride</u> (salt) is <u>one giant ionic lattice</u>, which is why salt crystals
   tend to be cuboid in shape. The <u>Na^+</u> and <u>Cl^-</u> <u>ions</u> are held together in a regular lattice.

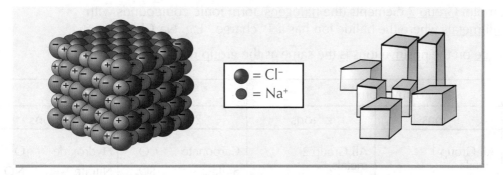

= Cl^-
= Na^+

Ionic Compounds All Have Similar Properties

1) They all have <u>high melting points</u> and <u>high boiling points</u> due to the <u>strong attraction</u>
   between the ions. It takes a large amount of <u>energy</u> to overcome this attraction.
   When ionic compounds <u>melt</u>, the ions are <u>free to move</u> and they'll <u>carry electric current</u>.

2) They do <u>dissolve easily</u> in water though. The ions <u>separate</u> and are
   all <u>free to move</u> in the solution, so they'll <u>carry electric current</u>.

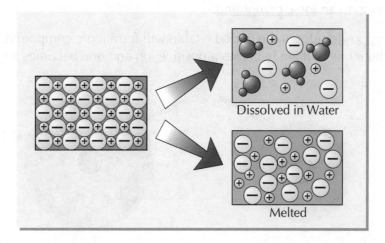

Dissolved in Water

Melted

Ionic compounds are always giant ionic lattices

You can get ionic compounds to <u>conduct electricity</u> by <u>melting</u> them or by <u>dissolving</u> them in water.
Dissolving them is easier though as it takes a lot of energy to melt an ionic compound.
Either way, it's the free ions that carry the electric current.

Ions

Make sure you've really got your head around the idea of ionic bonding before you start on this page.

Groups *1 & 2* and *6 & 7* are the Most Likely to Form *Ions*

1) Remember, atoms that have <u>lost</u> or <u>gained</u> an electron (or electrons) are <u>ions</u>.

2) Ions have the <u>electronic structure</u> of a <u>noble gas</u>.

3) The elements that most readily form ions are those in <u>Groups 1</u>, <u>2</u>, <u>6 and 7</u>.

4) <u>Group 1 and 2 elements</u> are <u>metals</u> and they <u>lose</u> electrons to form <u>positive ions</u>.

5) For example, <u>Group 1</u> elements (the <u>alkali metals</u>) form ionic compounds with <u>non-metals</u> where the metal ion has a 1$^+$ charge. E.g. K^+Cl^-.

6) <u>Group 6 and 7 elements</u> are <u>non-metals</u>. They <u>gain</u> electrons to form <u>negative ions</u>.

7) For example, <u>Group 7</u> elements (the <u>halogens</u>) form ionic compounds with the <u>alkali metals</u> where the halide ion has a 1$^-$ charge. E.g. Na^+Cl^-.

8) The <u>charge</u> on the <u>positive ions</u> is the <u>same</u> as the <u>group number</u> of the element:

Positive Ions		Negative Ions	
1$^+$ ions	2$^+$ ions	2$^-$ ions	1$^-$ ions
All Group 1 metals, including:	All Group 2 metals, including:	Carbonate CO_3^{2-} Sulfate SO_4^{2-}	Hydroxide OH^- Nitrate NO_3^-
Lithium Li^+ Sodium Na^+ Potassium K^+	Magnesium Mg^{2+} Calcium Ca^{2+}	All Group 6 elements, including: Oxide O^{2-} Sulfide S^{2-}	All Group 7 elements, including: Fluoride F^- Chloride Cl^- Bromide Br^- Iodide I^-

9) Any of the positive ions above can <u>combine</u> with any of the negative ions to form an <u>ionic compound</u>.

10) Only elements at <u>opposite sides</u> of the periodic table will form ionic compounds, e.g. Na and Cl, where one of them becomes a <u>positive ion</u> and one becomes a <u>negative ion</u>.

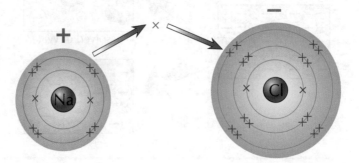

Remember, the + and – charges we talk about, e.g. Na$^+$ for sodium, just tell you <u>what type of ion the atom WILL FORM</u> in a chemical reaction. In sodium <u>metal</u> there are <u>only neutral sodium atoms, Na</u>. The Na$^+$ ions <u>will only appear</u> if the sodium metal <u>reacts</u> with something like water or chlorine.

Formulas of Ionic Compounds

Working out the <u>chemical formulas</u> for ionic compounds isn't too tricky if you know the <u>charges</u>.

Look at **Charges** to Work Out the **Formula** of an **Ionic Compound**

1) Ionic compounds are made up of a <u>positively charged</u> part and a <u>negatively charged</u> part.

2) The <u>overall charge</u> of <u>any compound</u> is <u>zero</u>.

3) So all the <u>negative charges</u> in the compound must <u>balance</u> all the <u>positive charges</u>.

4) You can use the charges on the <u>individual ions</u> present to work out the formula for the ionic compound.

Sodium Chloride

Sodium chloride contains Na^+ (+1) and Cl^- (–1) ions.
(+1) + (–1) = 0. The charges are balanced with one of each ion, so the formula for sodium chloride = NaCl.

NaCl

Magnesium Chloride

Magnesium chloride contains Mg^{2+} (+2) and Cl^- (–1) ions.
Because a chloride ion only has a 1⁻ charge we will need <u>two</u> of them to balance out the 2⁺ charge of a magnesium ion. This gives us the formula $MgCl_2$.

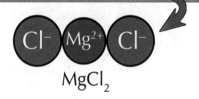

$MgCl_2$

The formula for exam success = revision...

The main thing to remember is that in compounds the <u>total charge must always add up to zero</u>.
So, for example, if you've got 2 positive charges, you need to balance it out with 2 negative charges.
Easy as that — as long as you know the <u>charges</u> of the ions involved, you can work out any formula.

Electronic Structure of Ions

This page has some lovely drawings of the <u>electronic structures</u> of ions.
As well as being colourful and pretty, they're really useful too.

*Show the Electronic Structure of **Simple** Ions with **Diagrams***

A useful way of representing ions is by <u>drawing</u> out their electronic structure. Just use a big <u>square bracket</u> and a + or − to show the charge. A few <u>ions</u> and the <u>ionic compounds</u> they form are shown below.

Sodium Chloride

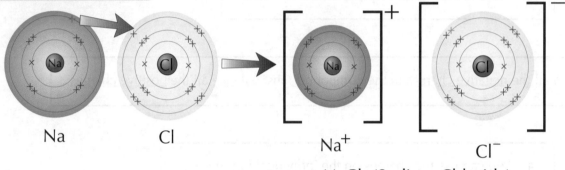

Na   Cl   Na$^+$ Cl$^-$

NaCl (Sodium Chloride)

Magnesium Oxide

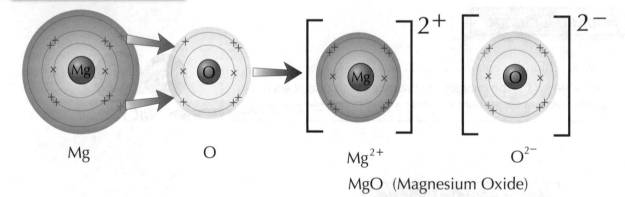

Mg   O   Mg^{2+} O^{2-}

MgO (Magnesium Oxide)

Calcium Chloride

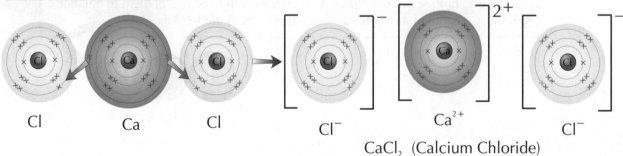

Cl   Ca   Cl   Cl$^-$ Ca^{2+} Cl$^-$

CaCl$_2$ (Calcium Chloride)

Show the electronic structure of ions with square brackets

Whether or not you're able to reproduce the drawings on this page all comes down to how well you've understood <u>ionic bonding</u>. (So if you're struggling, try reading the last few pages again.)

Warm-Up and Exam Questions

These questions will help you find out if you've learnt all the basics about ions and ionic bonding. Have a look back through the last few pages if you're unsure about any of these questions. It's tempting just to skip past anything you don't know, but it won't help you in the exam.

Warm-Up Questions

1) Sodium chloride has a giant ionic structure. Does it have a high or a low boiling point?
2) Why do ionic compounds conduct electricity when dissolved?
3) Do elements from Group 1 form positive ions or negative ions?
4) Do elements from Group 7 form positive ions or negative ions?
5) What is the formula of the compound containing Al^{3+} and OH^- ions only?

Exam Questions

1   The diagrams below show the electronic structures of sodium and fluorine.

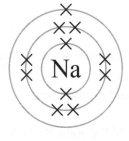

Sodium

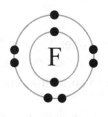

Fluorine

(a)   Describe what will happen when sodium and fluorine react, in terms of electrons.

(2 marks)

(b)   When sodium and fluorine react they form an ionic compound
Describe the structure of an ionic compound.

(3 marks)

2   The table below lists different ions.

Positive Ions		Negative Ions	
1⁺ ions	2⁺ ions	2⁻ ions	1⁻ ions
Lithium Li^+	Magnesium Mg^{2+}	Carbonate CO_3^{2-}	Hydroxide OH^-
		Sulfate SO_4^{2-}	Nitrate NO_3^-

Use the information in the table to write the formulas of the following ionic compounds:

(a)   magnesium carbonate.

(1 mark)

(b)   lithium sulfate.

(1 mark)

Exam Questions

3   When lithium reacts with oxygen it forms an ionic compound, Li_2O.
    (a)   Name the compound formed.

(1 mark)

    (b)   (i)   Complete the diagram below using arrows to show how the electrons
               are transferred when Li_2O is formed.

(1 mark)

          (ii)   Show the electron arrangements and the charges on the ions formed.

(2 marks)

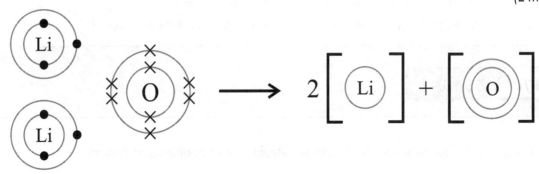

4   Magnesium (atomic number 12) and fluorine (atomic number 9) combine vigorously to form
    magnesium fluoride, an ionic compound.
    (a)   Draw a diagram to show the electron arrangement of each atom.

(2 marks)

    (b)   Give the symbol (including the charge) for each of the ions formed.

(2 marks)

    (c)   Using your answer to part (b), work out the formula of magnesium fluoride.

(1 mark)

    (d)   Once formed, explain why the ions remain together in a compound.

(1 mark)

    (e)   Magnesium fluoride has a giant ionic structure. Explain why:
          (i)   it doesn't melt easily.

(2 marks)

          (ii)   it conducts electricity when molten.

(1 mark)

5   Potassium and chlorine react to form potassium chloride.
    (a)   Complete the following table.

(2 marks)

	Potassium atom, K	Potassium ion, K^+	Chlorine atom, Cl	Chloride ion, Cl^-
Number of electrons	19			

    (b)   Draw a dot and cross diagram to show the formation of potassium chloride.

(2 marks)

Group 1 — The Alkali Metals

The alkali metals are <u>silvery solids</u> that have to be <u>stored in oil</u> and handled with <u>forceps</u> (they burn the skin).

Group 1 Metals are Known as the 'Alkali Metals'

1) Group 1 metals include <u>lithium</u>, <u>sodium</u> and <u>potassium</u>.

2) They all have <u>ONE outer shell electron</u>. This makes them <u>very reactive</u> and gives them all <u>similar properties</u>.

3) When the alkali metals react they all form <u>similar compounds</u> (see next page).

4) The alkali metals are <u>shiny</u> when freshly cut, but quickly react with the oxygen in <u>moist air</u> and <u>tarnish</u>.

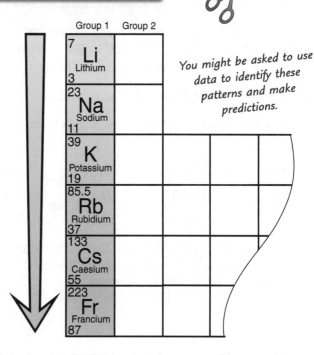

You might be asked to use data to identify these patterns and make predictions.

Reactivity Increases Down the Group

As you go <u>DOWN</u> Group 1, the alkali metals:

become <u>MORE REACTIVE</u> (see page 110)
...because the outer electron is <u>more easily lost</u>, because it's <u>further</u> from the nucleus.

have a <u>HIGHER DENSITY</u>
...because the atoms have <u>more mass</u>.

have a <u>LOWER MELTING POINT</u>

have a <u>LOWER BOILING POINT</u>

Group 1 — The Alkali Metals

Reaction with Cold Water Produces Hydrogen Gas

1) When lithium, sodium or potassium are put in water, they react very vigorously.

2) They move around the surface, fizzing furiously.

3) They produce hydrogen. Potassium gets hot enough to ignite it.
   If it hasn't already been ignited by the reaction, a lighted splint will indicate hydrogen by producing the notorious "squeaky pop" as it ignites.

4) The reaction makes an alkaline solution — this is why Group 1 is known as the alkali metals.

5) A hydroxide of the metal forms, e.g. sodium hydroxide (NaOH) or potassium hydroxide (KOH).

$$2Na_{(s)} + 2H_2O_{(l)} \rightarrow 2NaOH_{(aq)} + H_{2(g)}$$
$$2K_{(s)} + 2H_2O_{(l)} \rightarrow 2KOH_{(aq)} + H_{2(g)}$$

6) This experiment shows the relative reactivities of the alkali metals. The more violent the reaction, the more reactive the alkali metal is.

The Alkali Metals Form Ionic Compounds with Non-Metals

1) They are keen to lose their one outer electron to form a 1⁺ ion.

2) They are so keen to lose the outer electron there's no way they'd consider sharing, so covalent bonding is out of the question.

3) So they always form ionic bonds — and they produce white compounds that dissolve in water to form colourless solutions.

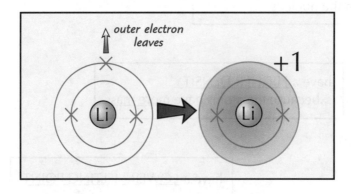

That reaction with water is the reason they're called alkali metals

The alkali metals all have very similar properties due to the fact that they've all got the same number of electrons in their outer shell. This means that they all react in the same way, for example, with water and with non-metals. Once you know about one alkali metal, you know about them all.

Group 7 — The Halogens

The 'trend thing' happens in Group 7 as well — that shouldn't come as a surprise.
But some of the trends are kind of the opposite of the Group 1 trends.

Reactivity Decreases Down the Group

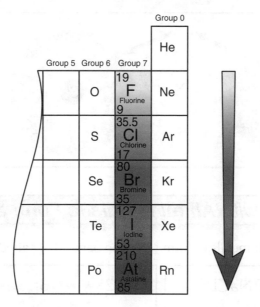

As you go <u>DOWN</u> Group 7, the <u>HALOGENS</u> have the following properties:

1) <u>LESS REACTIVE</u>
   ...because it's <u>harder to gain</u> an extra electron, because the outer shell's <u>further</u> from the nucleus.

2) <u>HIGHER MELTING POINT</u>

3) <u>HIGHER BOILING POINT</u>

The **Halogens** are all **Non-metals** with **Coloured Vapours**

<u>Fluorine</u> is a very reactive, poisonous <u>yellow gas</u>.

<u>Chlorine</u> is a fairly reactive, poisonous <u>dense green gas</u>.

<u>Bromine</u> is a dense, poisonous, <u>red-brown volatile liquid</u>.

<u>Iodine</u> is a <u>dark grey</u> crystalline <u>solid</u> or a <u>purple vapour</u>.

They all exist as molecules which are <u>pairs of atoms</u>:

F_2  Cl_2  Br_2  I_2

Group 7 — The Halogens

The Halogens Form **Ionic Bonds** with **Metals**

The halogens form 1⁻ ions called halides (F⁻, Cl⁻, Br⁻ and I⁻)
when they bond with metals, for example Na^+Cl^- or $Fe^{3+}Br^-_3$.
The diagram shows the bonding in sodium chloride, NaCl.

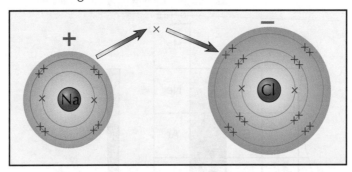

The Halogens React with **Alkali Metals** to Form **Salts**

They react vigorously with alkali metals to form salts (see page 110) called 'metal halides'.

$$2Na + Cl_2 \rightarrow 2NaCl$$
sodium + chlorine → sodium chloride

$$2K + Br_2 \rightarrow 2KBr$$
potassium + bromine → potassium bromide

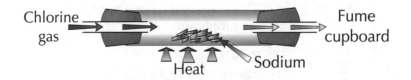

Chlorine gas   Fume cupboard

Heat   Sodium

More Reactive Halogens Will **Displace** Less Reactive Ones

A more reactive halogen can displace (kick out) a less reactive halogen
from an aqueous solution of its salt.

E.g. chlorine can displace bromine and iodine from an aqueous solution of its salt
(a bromide or iodide). Bromine will also displace iodine because of the trend in reactivity.

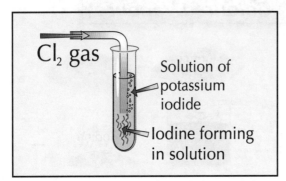

Cl₂ gas

Solution of
potassium
iodide

Iodine forming
in solution

$$Cl_{2\,(g)} + 2KI_{\,(aq)} \rightarrow I_{2\,(aq)} + 2KCl_{\,(aq)}$$

$$Cl_{2\,(g)} + 2KBr_{\,(aq)} \rightarrow Br_{2\,(aq)} + 2KCl_{\,(aq)}$$

Polish that halo and get revising...

Displacement reactions needn't be too confusing — you just have to figure out which halogen's the
most reactive. If a halogen is higher up the group than another one, it'll kick that one out of the
solution and get in there itself. Just like how I displace my sister from the sofa before Corrie.

Group 0 — The Noble Gases

The <u>noble gases</u> — stuffed full of every honourable virtue. They <u>don't react</u> with very much and you can't even see them — making them, well, a bit <u>dull</u> really.

Group 0 Elements are All **Inert**, **Colourless Gases**

1) Group 0 elements are called the <u>noble gases</u> and include the elements <u>helium</u>, <u>neon</u> and <u>argon</u> (plus a few others).

2) All elements in Group 0 are <u>colourless gases</u> at room temperature.

3) They are also more or less <u>inert</u> — this means they <u>don't react</u> with much at all. The reason for this is that they have a <u>full outer shell</u>. This means they're <u>not</u> desperate to <u>give up</u> or <u>gain</u> electrons.

4) As the noble gases are inert they're <u>non-flammable</u> — they won't set on fire.

	Group 6	Group 7	Group 0
			4 **He** Helium 2
	O	F	20 **Ne** Neon 10
	S	Cl	40 **Ar** Argon 18
	Se	Br	84 **Kr** Krypton 36
	Te	I	131 **Xe** Xenon 54
	Po	At	222 **Rn** Radon 86

Noble Gases Took a While to **Discover**

1) Their properties make the gases <u>hard to observe</u> — it took a long time for them to be discovered.

2) The gases were found when chemists noticed that the <u>density</u> of <u>nitrogen</u> made in chemical <u>reactions</u> was <u>different</u> to the density of nitrogen taken from the <u>air</u>.

3) They hypothesised that the nitrogen obtained from air must have <u>other gases mixed in with it</u>.

4) Scientists gradually discovered the different noble gases through a series of <u>experiments</u>, including the <u>fractional distillation</u> of <u>air</u> (see page 88).

Group 0 — The Noble Gases

The Noble Gases have **Many Everyday Uses**...

Argon

1) <u>Argon</u> is used to provide an <u>inert atmosphere</u> in <u>filament lamps</u> (light bulbs).
2) As the argon is <u>non-flammable</u> it stops the very hot filament from <u>burning away</u>.
3) It can also be used to protect metals that are being <u>welded</u>.
4) The inert atmosphere stops the hot metal reacting with <u>oxygen</u>.

Helium

1) <u>Helium</u> is used in <u>airships</u> and <u>party balloons</u>.
2) Helium has a <u>lower density</u> than air — so it makes balloons <u>float</u>.

There are **Patterns** in the **Properties** of the Noble Gases

The <u>boiling points</u> and <u>densities</u> of the noble gases <u>increase</u> as you move <u>down</u> the group.

Noble Gas	Boiling Point (°C)	Density (g/cm³)
helium	−269	0.0002
neon	−246	0.0009
argon	−186	0.0018
krypton	−153	0.0037
xenon	−108	0.0059
radon	−62	0.0097

Just like other groups of elements, the noble gases follow patterns...

The noble gases might be a bit boring, but there's nothing wrong with that. Although they're unreactive and hard to see, they're actually pretty useful. It took a while to discover them, but we now know all about the trends in their properties, like the patterns in their densities and boiling points.

Warm-Up and Exam Questions

These questions are all about the groups of the periodic table that you need to know about. Treat the exam questions like the real thing — don't look back through the book until you've finished.

Warm-Up Questions

1) In Group 1, as you go down the periodic table, does the reactivity increase or decrease?
2) Which gas is produced when an alkali metal reacts with water?
3) In Group 7, what is the trend in physical state as you go down the group?
4) Give an example of a salt produced when a Group 1 metal reacts with a Group 7 element.
5) Write down the balanced equation for the displacement of iodine from potassium iodide by bromine.
6) Why did it take a long time for the noble gases to be discovered?

Exam Questions

1   Chlorine is a Group 7 element used in water purification.
    Its electron arrangement is shown in the diagram below.

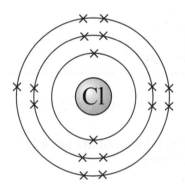

(a)   Chlorine can react with Group 1 metals to form ionic compounds.
      Draw a diagram to show the electron arrangement of a chloride ion.

                                                                        (1 mark)

(b)   Chlorine is more reactive than bromine.
      When chlorine is bubbled through potassium bromide solution a reaction occurs.

   (i)   Write a word equation for the reaction.

                                                                        (1 mark)

   (ii) State the colour that the solution will be after the reaction.

                                                                        (1 mark)

   (iii) Explain why chlorine atoms are more reactive than bromine atoms.

                                                                        (2 marks)

Exam Questions

2   All the Group 1 metals react vigorously with water.

   (a) Explain why this is.

   (1 mark)

   (b) The Group 1 metals all react with water at a different rate. Explain why this is.

   (2 marks)

   (c) Give the two products formed in these reactions.

   (2 marks)

   (d) During these reactions a solution is formed.
        Is this solution acidic, neutral or alkaline?

   (1 mark)

3   The table shows some of the physical properties of four of the halogens.

Halogen	Atomic number	Colour	Physical state at room temperature	Boiling point
			Properties	
Fluorine	9	yellow		−188 °C
Chlorine	17	green		−34 °C
Bromine	35	red-brown		59 °C
Iodine	53	dark grey		185 °C

   (a) Give the physical state at room temperature of all four halogens.

   (4 marks)

   (b) Draw an arrow next to the left hand side of the table to show the direction of
        increasing reactivity in the halogens.

   (1 mark)

   (c) This equation shows a reaction between chlorine and potassium iodide.

   $$Cl_2(g) + 2KI(aq) \rightarrow I_2(aq) + 2KCl(aq)$$

   (i)   What type of reaction is this?

   (1 mark)

   (ii) Which is the less reactive halogen in this reaction?

   (1 mark)

4   Helium is an inert, colourless gas. It is one of the noble gases.

   (a) State the Group number of helium.

   (1 mark)

   (b) Describe the trend in the boiling points of the noble gases as you move
        down the Group.

   (1 mark)

Covalent Bonding

Some elements bond ionically (see page 102) but others form strong <u>covalent bonds</u>.
This is where atoms <u>share electrons</u> with each other so that they've got <u>full outer shells</u>.

Covalent Bonds — Sharing Electrons

1) Sometimes atoms prefer to make <u>covalent bonds</u> by <u>sharing</u> electrons with other atoms.

2) They only share electrons in their <u>outer shells</u> (highest energy levels).

3) This way <u>both</u> atoms feel that they have a <u>full outer shell</u>, and that makes them happy. Having a full outer shell gives them the electronic structure of a <u>noble gas</u>.

4) Each <u>covalent bond</u> provides one <u>extra</u> shared electron for each atom.

5) So, a covalent bond is a <u>shared pair</u> of electrons.

6) Each atom involved has to make <u>enough</u> covalent bonds to <u>fill up</u> its outer shell.

7) Here are some <u>important examples</u>:

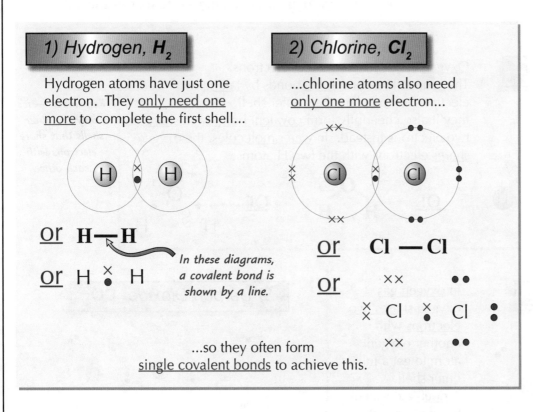

1) Hydrogen, **H₂**

Hydrogen atoms have just one electron. They <u>only need one more</u> to complete the first shell...

In these diagrams, a covalent bond is shown by a line.

...so they often form <u>single covalent bonds</u> to achieve this.

2) Chlorine, **Cl₂**

...chlorine atoms also need <u>only one more</u> electron...

These dot and cross diagrams only show the outer shell of electrons.

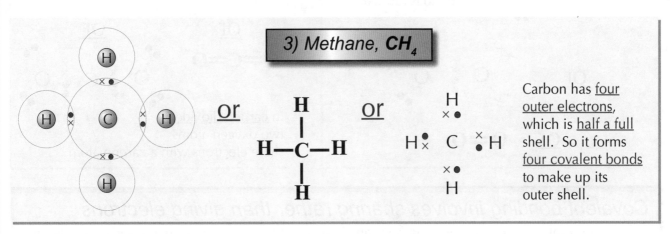

3) Methane, **CH₄**

Carbon has <u>four outer electrons</u>, which is <u>half a full</u> shell. So it forms <u>four covalent bonds</u> to make up its outer shell.

More Covalent Bonding

There are five more examples of covalent bonding on this page — just a few diagrams and a smattering of words. What a pleasant page.

4) Hydrogen Chloride, **HCl**

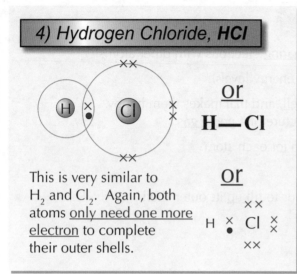

or

H—Cl

This is very similar to H_2 and Cl_2. Again, both atoms only need one more electron to complete their outer shells.

or

H ×•× Cl ×××

5) Ammonia, **NH₃**

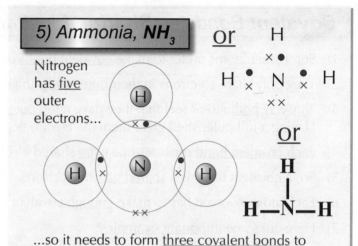

or

H ×• N •× H
   ××

or

H
|
H—N—H

...so it needs to form three covalent bonds to make up the extra three electrons needed.

Nitrogen has five outer electrons...

6) Water, **H₂O**

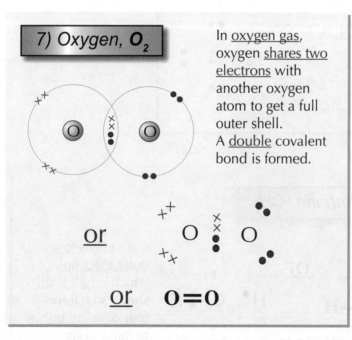

Oxygen atoms have six outer electrons. They sometimes form ionic bonds by taking two electrons to complete their outer shell. However they'll also cheerfully form covalent bonds and share two electrons instead. In water molecules, the oxygen shares electrons with the two H atoms.

or

H   **O**   **H**

or

O
×• ×•
H   H

Remember — it's only the outer shells that share electrons with each other.

7) Oxygen, **O₂**

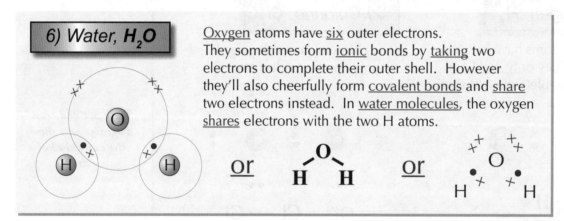

In oxygen gas, oxygen shares two electrons with another oxygen atom to get a full outer shell. A double covalent bond is formed.

or

O ×•× O

or   **O=O**

8) Carbon Dioxide, **CO₂**

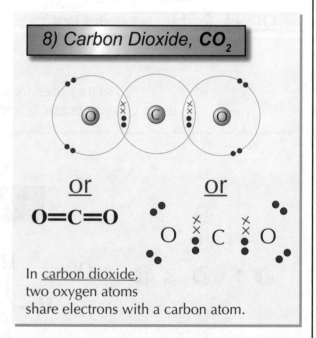

or

O=C=O

or

O ××• C •×× O

In carbon dioxide, two oxygen atoms share electrons with a carbon atom.

Covalent bonding involves sharing rather than giving electrons

Every atom wants a full outer shell, and they can get that either by becoming an ion (see page 104) or by sharing electrons. Some atoms can share two electron pairs — it's not limited to a single bond.

Covalent Substances

Substances with <u>covalent bonds</u> (electron sharing) can form <u>simple molecules</u> or <u>giant structures</u>.

Simple *Molecular* Substances

1) The atoms form <u>very strong</u> covalent bonds to form <u>small</u> molecules of several atoms.

2) By contrast, the forces of attraction <u>between</u> these molecules are <u>very weak</u>.

3) The result of these feeble <u>intermolecular forces</u> is that the <u>melting</u> and <u>boiling points</u> are <u>very low</u>, because the molecules are <u>easily parted</u> from each other. It's the <u>intermolecular forces</u> that get <u>broken</u> when simple molecular substances melt or boil — <u>not</u> the much <u>stronger covalent bonds</u>.

4) Most molecular substances are <u>gases or liquids</u> at room temperature, but they can be <u>solids</u>.

5) Molecular substances <u>don't conduct electricity</u> — there are <u>no ions</u> so there's <u>no electrical charge</u>.

Very weak intermolecular forces

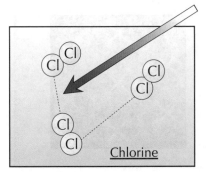

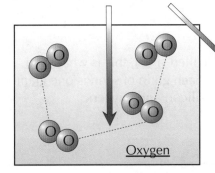

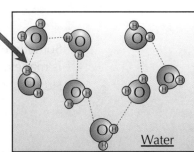

Chlorine   Oxygen   Water

Giant Covalent Structures Are *Macromolecules*

1) These are similar to giant ionic structures (lattices) <u>except</u> that there are <u>no charged ions</u>.

2) <u>All</u> the atoms are <u>bonded</u> to <u>each other</u> by <u>strong</u> covalent bonds.

3) This means that they have <u>very high</u> melting and boiling points.

4) They <u>don't conduct electricity</u> — not even when <u>molten</u> (except for graphite).

5) The <u>main examples</u> are <u>diamond</u> and <u>graphite</u>, which are both made only from <u>carbon atoms</u>, and <u>silicon dioxide</u> (silica) — see the next page.

Covalent Substances

Here are three examples of giant covalent structures.

Diamond

Each carbon atom forms four covalent bonds in a very rigid giant covalent structure. This structure makes diamond the hardest natural substance, so it's used for drill tips. And it's pretty and sparkly too.

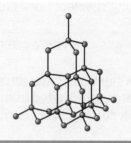

Silicon Dioxide (Silica)

Sometimes called silica, this is what sand is made of. Each grain of sand is one giant structure of silicon and oxygen.

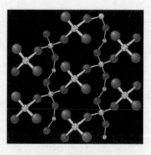

Graphite

Each carbon atom only forms three covalent bonds. This creates layers which are free to slide over each other, like a pack of cards — so graphite is soft and slippery.

The layers are held together so loosely that they can be rubbed off onto paper — that's how a pencil works. This is because there are weak intermolecular forces between the layers.

Graphite is the only non-metal which is a good conductor of heat and electricity. Each carbon atom has one delocalised (free) electron and it's these free electrons that conduct heat and electricity.

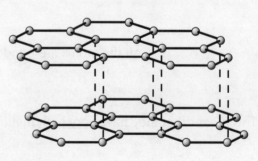

Graphite and diamond contain exactly the same atoms

Graphite and diamond are both made purely from carbon — there's no difference at all in their atoms. The difference in properties (and price) of the two substances is all down to the way the atoms are held together. Don't get confused though — they're both still giant covalent substances.
Giant covalent substances and simple molecular substances have very different properties because they have very different structures. Look back at the diagrams showing their bonding if you need a reminder.

Metallic Structures

Ever wondered what makes <u>metals</u> tick? Well, either way, this is the page for you.

Metal Properties Are All Due to the Sea of Free Electrons

1) <u>Metals</u> also consist of a <u>giant structure</u>.

2) <u>Metallic bonds</u> involve the all-important '<u>free electrons</u>' which produce <u>all</u> the properties of metals. These delocalised (free) electrons come from the <u>outer shell</u> of <u>every</u> metal atom in the structure.

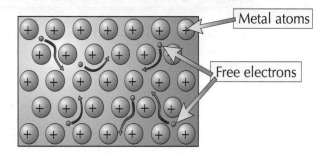

3) These electrons are <u>free to move</u> through the whole structure and so metals are good conductors of <u>heat and electricity</u>.

4) These electrons also <u>hold</u> the <u>atoms</u> together in a <u>regular</u> structure. There are strong forces of <u>electrostatic attraction</u> between the <u>positive metal ions</u> and the <u>negative electrons</u>.

5) They also allow the layers of atoms to <u>slide</u> over each other, allowing metals to be <u>bent</u> and <u>shaped</u>.

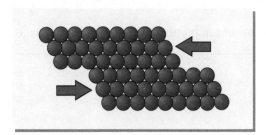

Alloys are Harder Than Pure Metals

1) <u>Pure metals</u> often aren't quite right for certain jobs. So scientists <u>mix two or more metals together</u> — creating an <u>alloy</u> with the properties they want.

2) Different elements have <u>different sized atoms</u>. So when another metal is mixed with a pure metal, the new metal atoms will <u>distort</u> the layers of metal atoms, making it more difficult for them to slide over each other. So alloys are <u>harder</u>.

Identifying Structures

If you know the <u>properties</u> of the four types of substance in this section then this page shouldn't be a problem. If you don't then it's worth taking another look at the last few pages.

Identifying the Structure of a Substance by Its Properties

You can <u>identify</u> most substances just by the way they <u>behave</u> as either:

- <u>giant ionic</u>,
- <u>simple molecular</u>,
- <u>giant covalent</u>,
- or <u>giant metallic</u>.

<u>Example</u>:

Four substances were tested for various properties with the following results.

Identify the structure of each substance. (Answers on page 256.)

Substance	Melting point (°C)	Boiling point (°C)	Good electrical conductor?
A	−219.62	−188.12	No
B	1535	2750	Yes
C	1410	2355	No
D	770	1420	When molten

Look at the properties to identify the structure

If you know the properties of a substance then you can work out its structure — and explain why it has those properties, of course. It's not as hard as it sounds because a substance will always be one of four things — giant ionic, simple molecular, giant covalent or giant metallic. So, if you know how to identify those four substances, then you know how to identify anything. Phew.

Warm-Up and Exam Questions

Don't charge past this page, it's a lot more important than it looks.

Warm-Up Questions

1) How is covalent bonding different from ionic bonding?
2) Why does chlorine have a very low boiling point?
3) Describe the differences in the physical properties of diamond and graphite.
4) Give another example of a substance that has a giant covalent structure.

Exam Questions

1 Methane is a covalently bonded molecule with the formula CH_4.
Draw a dot and cross diagram for the methane molecule,
showing only the outer electrons.

(2 marks)

2 The table compares some physical properties of silicon dioxide, bromine and graphite.

Property	silicon dioxide	bromine	graphite
Melting point (°C)	1610	−7	3657
Electrical conductivity	poor	poor	good
Solubility in water	insoluble	slightly soluble	insoluble

(a) What is the structure of:
  (i) silicon dioxide?

(1 mark)

  (ii) graphite?

(1 mark)

  (iii) bromine?

(1 mark)

(b) Explain why bromine has poor electrical conductivity.

(1 mark)

(c) Explain why graphite has good electrical conductivity.

(1 mark)

(d) Bromine is a liquid at room temperature (20 °C).
Explain why bromine has such a low melting point compared with
silicon dioxide and graphite.

(2 marks)

Exam Questions

3   The diagram below shows the arrangement of atoms in pure iron.

Steel is an alloy of iron and carbon.
(a) Draw a similar diagram to show the arrangement of atoms in steel.

(2 marks)

(b) Steel is harder than iron. Explain why.

(3 marks)

4   The table gives data for some physical properties of a selection of substances.

Substance	Melting point (°C)	Boiling point (°C)	Electrical conductivity
A	-219	-183	poor
B	3550	4827	poor
C	1495	2870	good
D	801	1413	good when molten

(a) In what state would you expect substance D to be at room temperature?

(1 mark)

(b) What is the structure of:
(i)   substance B?

(1 mark)

(ii) substance D?

(1 mark)

(c) Substance A is oxygen.
(i)   Draw a dot and cross diagram to show the outer electrons in an oxygen molecule.

(2 marks)

(ii) Explain why oxygen has such a low melting point.

(2 marks)

(d) Substance C is a metal. According to the table, it is a good conductor of electricity.
(i)   Explain why this is.

(2 marks)

(ii) Would substance C be a good conductor of heat?

(1 mark)

Superconductors

If you make them <u>cold</u> enough, metals can start behaving in a pretty odd way.

At **Very Low Temperatures**, Some Metals are **Superconductors**

1) Normally, all metals have some <u>electrical resistance</u>.

2) That resistance means that whenever electricity flows through them, they <u>heat up</u>, and some of the <u>electrical energy</u> is <u>wasted</u> as <u>heat</u>.

3) If you make some metals <u>cold</u> enough, though, their <u>resistance disappears completely</u>. The metal becomes a <u>superconductor</u>.

4) Without any <u>resistance</u>, none of the <u>electrical energy</u> is turned into <u>heat</u>, so none of it's <u>wasted</u>.

5) That means you could start a <u>current flowing</u> through a <u>superconducting circuit</u>, take out the <u>battery</u>, and the current would carry on flowing <u>forever</u>.

So What's the **Catch**...

1) Using <u>superconducting wires</u> you can make:

   a) <u>Power cables</u> that transmit electricity without any loss of power (loss-free power transmission).

   b) Really <u>strong electromagnets</u> that don't need a constant power source.

   c) <u>Electronic circuits</u> that work really fast, because there's no resistance to slow them down.

2) But they need to be <u>REALLY COLD</u>. Metals only start superconducting at <u>less than −265 °C</u>! Getting things that cold is <u>very hard</u>, and <u>very expensive</u>, which limits the use of superconductors.

3) Scientists are trying to develop <u>room temperature</u> superconductors now. So far, they've managed to get some weird <u>metal oxide</u> things to superconduct at about <u>−135 °C</u>, which is a <u>much cheaper</u> temperature to get down to. But ideally they need to develop superconductors that work at <u>20 °C</u>.

Superconductors have various uses...

... for example, superconducting magnets are used in magnetic resonance imaging (MRI) scanners in hospitals. The <u>huge</u> magnetic fields they need can be generated without using up a load of electricity.

Transition Metals

The middle section of the periodic table is made up of some elements known as <u>transition metals</u>.

Metals in the Middle of the Periodic Table are Transition Metals

1) A lot of everyday metals are transition metals (e.g. copper, iron, zinc, gold, silver, platinum) — but there are <u>lots</u> of others as well.
2) Transition metals have typical 'metallic' properties.
3) If you get asked about a transition metal you've never heard of — <u>don't panic</u>.
   These 'new' transition metals follow <u>all</u> the properties you've <u>already learnt</u> for the others.

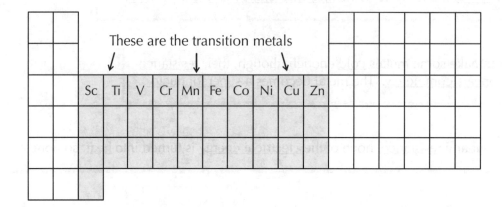

These are the transition metals

| | | Sc | Ti | V | Cr | Mn | Fe | Co | Ni | Cu | Zn | | | | | | |

Transition metals and their compounds make good catalysts

1) <u>Iron</u> is the catalyst used in the <u>Haber process</u> for making <u>ammonia</u>.
2) <u>Nickel</u> is useful for the <u>hydrogenation</u> of alkenes (e.g. to make margarine).

Transition metals often have more than one ion, e.g. Fe^{2+}, Fe^{3+}

Two other examples are <u>copper</u>: Cu^+ and Cu^{2+} and <u>chromium</u>: Cr^{2+} and Cr^{3+}.

The compounds are very colourful

The compounds of transition elements are colourful due to the <u>transition metal ion</u> they contain. E.g. Iron(II) compounds are usually <u>light green</u>, iron(III) compounds are <u>orange/brown</u> (e.g. rust) and copper compounds are often <u>blue</u>.

Mendeleev and his amazing technicoloured periodic table...

There are lots of different sections of elements in the periodic table — for example, the alkali metals, the halogens, etc. Like these groups, the transition metals all have lots of properties in common.

Thermal Decomposition

It's your lucky day — there are <u>TWO</u> exciting <u>reactions</u> coming up over the next two pages.

Thermal Decomposition — *Breaking* Down with *Heat*

1) <u>Thermal decomposition</u> is when a substance <u>breaks down</u> into at least two other substances when <u>heated</u>.

2) <u>Transition metal carbonates</u> break down on heating. Transition metal carbonates are things like copper(II) carbonate ($CuCO_3$), iron(II) carbonate ($FeCO_3$), zinc carbonate ($ZnCO_3$) and manganese carbonate ($MnCO_3$), i.e. they've all got a <u>CO_3</u> bit in them.

3) They break down into a <u>metal oxide</u> (e.g. copper oxide, CuO) and <u>carbon dioxide</u>. This usually results in a <u>colour change</u>.

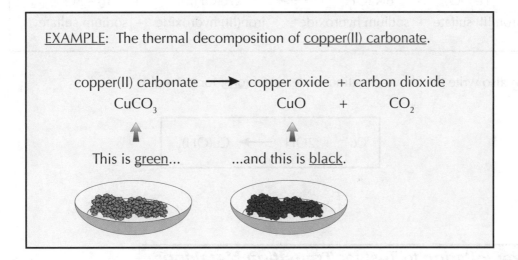

<u>EXAMPLE</u>: The thermal decomposition of <u>copper(II) carbonate</u>.

copper(II) carbonate $\longrightarrow$ copper oxide + carbon dioxide
$CuCO_3$ $\qquad$ CuO $\quad + \quad$ CO_2

This is <u>green</u>... $\qquad$ ...and this is <u>black</u>.

The reactions for the thermal decomposition of:

(i) <u>iron(II) carbonate</u> to iron oxide (FeO),

(ii) <u>manganese carbonate</u> to manganese oxide (MnO),

(iii) <u>zinc carbonate</u> to zinc oxide (ZnO),

are the same — although the colours are different.

4) You can easily check that the gas given off is <u>carbon dioxide</u> by bubbling it through <u>limewater</u>. If carbon dioxide is present the limewater turns <u>cloudy</u>.

Precipitation

Precipitation — A *Solid* Forms in *Solution*

1) A <u>precipitation</u> reaction is where <u>two solutions</u> react and an insoluble <u>solid</u> forms in the solution.

2) The solid is said to '<u>precipitate out</u>' and, confusingly, the solid is also called '<u>a precipitate</u>'.

3) Some soluble <u>transition metal compounds</u> react with <u>sodium hydroxide</u> to form an <u>insoluble hydroxide</u>, which then precipitates out. Here are some examples:

$$CuSO_4 \quad + \quad 2NaOH \quad \longrightarrow \quad Cu(OH)_2 \quad + \quad Na_2SO_4$$
copper(II) sulfate + sodium hydroxide   copper(II) hydroxide + sodium sulfate

$$FeSO_4 \quad + \quad 2NaOH \quad \longrightarrow \quad Fe(OH)_2 \quad + \quad Na_2SO_4$$
iron(II) sulfate + sodium hydroxide   iron(II) hydroxide + sodium sulfate

$$Fe_2(SO_4)_3 \quad + \quad 6NaOH \quad \longrightarrow \quad 2Fe(OH)_3 \quad + \quad 3Na_2SO_4$$
iron(III) sulfate + sodium hydroxide   iron(III) hydroxide + sodium sulfate

4) You can also write the above equations in terms of <u>ions</u>, for example:

$$Cu^{2+} + 2OH^- \longrightarrow Cu(OH)_2$$

Use *Precipitation* to Test for *Transition Metal Ions*

1) Some insoluble <u>transition metal hydroxides</u> have distinctive <u>colours</u>.

Copper(II) hydroxide is a <u>blue</u> solid.
Iron(II) hydroxide is a <u>grey/green</u> solid.
Iron(III) hydroxide is an <u>orange/brown</u> solid.

2) You can use this fact to <u>test</u> which transition metal ions a solution contains.

3) For example, if you add sodium hydroxide to an <u>unknown soluble salt</u>, and an <u>orange/brown</u> precipitate forms, you know you've got iron(III) hydroxide and so have <u>Fe</u>$^{3+}$ ions in the solution.

Precipitation reactions — put up an umbrella...

<u>Bad news</u>: there's some tricky chemistry to get your head around on this page. <u>Good news</u>: it's the <u>end</u> of a section. But before you get a well earned biccie, have a go at the <u>questions</u> on the <u>next two pages</u>.

Warm-Up and Exam Questions

Warm-Up Questions

1) Give one disadvantage of using superconductors.
2) Give three typical properties of transition metals.
3) What name is given to a reaction in which a substance is broken down by heating?
4) What name is given to a reaction in which two solutions react to give an insoluble salt?

Exam Questions

1   The table below shows some of the properties of four common metals.

Metal	Melting point (°C)	Density (g/cm³)	Effect of heating in air	Ions formed
Magnesium	650	1.74	Burns very brightly	Mg^{2+}
Iron	1535	7.87	Produces sparks if powdered	Fe^{2+}, Fe^{3+}
Chromium	1860	7.19	Little reaction	Cr^{2+}, Cr^{3+}
Sodium	98	0.97	Burns very brightly	Na^+

(a) The properties of iron and chromium are typical of transition metals.
The properties of magnesium and sodium are typical of Group 1 and 2 metals.
Suggest three differences between the two types of metal.

(3 marks)

(b) Give two further characteristic properties that distinguish transition metals from other metals.

(2 marks)

(c) Iron(II) carbonate will thermally decompose.
Carbon dioxide is produced in this reaction.

(i)   Describe how you can test for the presence of carbon dioxide.

(1 mark)

(ii) Write a word equation for the thermal decomposition of iron(II) carbonate.

(1 mark)

2   The critical temperature (T_c) at which some materials become superconducting is shown in the table.

Type	T_c (°C)
zinc	−272
aluminum	−272
tin	−269
mercury	−269
metal oxide ceramic 1	−145
metal oxide ceramic 2	−123

(a) What does 'superconducting' mean?

(2 marks)

(b) Which of the materials in the table is most likely to be useful in real-life applications of superconductors? Explain your answer.

(1 mark)

(c) Give two potential applications of superconducting materials.

(2 marks)

Revision Summary for Section 6

And just when you realise that you can't take any more Chemistry, it's the end of the section. Phew. Which means it's time for some more questions. If you can't answer any of them, look back in the book.

1) What size groups did Döbereiner organise the elements into? What were these groups called?

2) Give two reasons why Newlands' Octaves were criticised.

3) Why did Mendeleev leave gaps in his Table of Elements?

4) How are the group number and the number of electrons in the outer shell related?

5) What is shielding?

6)* The table on the right gives the masses and relative abundances of the isotopes of neon. Calculate the relative atomic mass of neon. Give your answer to 2 decimal places.

relative mass of isotope	relative abundance
20	91%
22	9%

7) Describe the structure of a crystal of sodium chloride.

8) List the main properties of ionic compounds.

9) Use information from the periodic table to help you work out the formulas of these ionic compounds:
a) potassium chloride   b) calcium chloride

10) Which group are the alkali metals?

11) Describe the trend in reactivity as you go down the group of alkali metals.

12) Explain why Group 7 elements get less reactive as you go down the group from fluorine to iodine.

13) Will the following reactions occur:
a) iodine with lithium chloride, b) chlorine with lithium bromide?

14) Explain why the noble gases are inert.

15) Name two noble gases and state a use for each.

16) Sketch dot and cross diagrams showing the bonding in molecules of:
a) hydrogen,   b) hydrogen chloride,   c) water,   d) ammonia

17) What are the two types of covalent substance? Give three examples of each.

18) List three properties of metals and explain how metallic bonding causes these properties.

19) Explain why alloys are harder than pure metals.

20)* Identify the structure of each of the substances in the table:

Substance	Melting point (°C)	Electrical conductivity	Hardness [scale of 0 – 10 (10 being diamond)]
A	3410	Very high	7.5
B	2072	Zero	9
C	605	Zero in solid form. High when molten.	Low

21) Explain the advantage of using a superconductor rather than a normal wire to make a circuit.

22) Name six transition metals, and give uses for two of them.

23) What are thermal decomposition reactions?

24) What type of reaction between two liquids results in the formation of a solid? What are these solid products called?

* Answers on page 257

Relative Formula Mass

The biggest trouble with <u>relative atomic mass</u> and <u>relative formula mass</u> is that they <u>sound</u> so blood-curdling. Take a few deep breaths, and just enjoy, as the mists slowly clear...

Relative Atomic Mass, A_r, is Easy

1) You'll probably remember about <u>relative atomic mass</u> A_r from page 100.
2) The A_r for each element is the <u>same</u> as its <u>mass number</u> (see page 16).
3) So, you can find it easily by looking at the <u>periodic table</u> — the <u>bigger number</u> for each element is the <u>relative atomic mass</u>. For example:

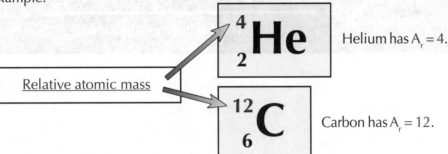

Helium has $A_r = 4$.

Carbon has $A_r = 12$.

Relative atomic mass

Relative Formula Mass, M_r, is Also Easy

If you have a compound like $MgCl_2$ then it has a <u>relative formula mass</u>, M_r, which is just all the relative atomic masses <u>added together</u>.

For $MgCl_2$ it would be:

The relative atomic mass of chlorine is multiplied by 2 because there are two chlorine atoms.

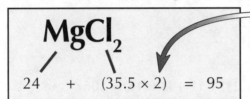

$$24 + (35.5 \times 2) = 95$$

So M_r for $MgCl_2$ is simply <u>95</u>.

You can easily get A_r for any element from the periodic table, but in a lot of questions they give you them anyway. And that's all it is. A big fancy name like <u>relative formula mass</u> and all it means is "<u>add up all the relative atomic masses</u>".

"ONE MOLE" of a Substance is Equal to its M_r in Grams

The <u>relative formula mass</u> (A_r or M_r) of a substance <u>in grams</u> is known as <u>one mole</u> of that substance.

<u>Examples:</u>

Iron has an A_r of 56.

Nitrogen gas, N_2, has an M_r of 28 (2 × 14).

So one mole of iron weighs exactly 56 g

So one mole of N_2 weighs exactly 28 g

You can convert between moles and grams using this formula:

<u>NUMBER OF MOLES</u> = <u>Mass in g (of element or compound)</u>
M_r (of element or compound)

<u>Example:</u> How many moles are there in 42 g of carbon?

<u>Answer:</u> No. of moles = Mass (g) / M_r = 42/12 = <u>3.5 moles</u>   Easy Peasy

Formula Mass Calculations

Although relative atomic mass and relative formula mass are <u>easy enough</u>, it can get just a tad <u>trickier</u> when you start getting into other calculations which use them. It depends on how good your maths is basically, because it's all to do with ratios and percentages.

Calculating % Mass of an Element in a Compound

This is actually dead easy — so long as you've learnt this formula:

$$\text{Percentage mass of an element in a compound} = \frac{A_r \times \text{no. of atoms (of that element)}}{M_r \text{ (of whole compound)}} \times 100$$

Example:

Find the percentage mass of sodium in sodium carbonate, Na_2CO_3.

ANSWER:

- A_r of sodium = 23
- A_r of carbon = 12
- A_r of oxygen = 16

M_r of Na_2CO_3 = $(2 \times 23) + 12 + (3 \times 16) = 106$

Now use the formula:

$$\text{Percentage Mass} = \frac{A_r \times \text{No. of atoms}}{M_r} \times 100 = \frac{23 \times 2}{106} \times 100 = 43.4\%$$

And there you have it.
Sodium makes up <u>43.4%</u> of the mass of sodium carbonate.

You can't just read these pages — work through the examples too

As usual with these calculations, <u>practice makes perfect</u>. Try these:

Find the percentage mass of oxygen in each of these:

Don't skip this bit, you'll be glad you're perfect when it comes to exam day.

a) Fe_2O_3
b) H_2O
c) $CaCO_3$
d) H_2SO_4

Answers on page 257.

Empirical Formulas and Molar Volume

Finding the **Empirical Formula** (from Masses or Percentages)

This also sounds a lot worse than it really is. Try this for a nice simple stepwise method:

1) List all the elements in the compound (there's usually only two or three).

2) Underneath them, write their experimental masses or percentages.

3) Divide each mass or percentage by the A_r for that particular element.

4) Turn the numbers you get into a nice simple ratio by multiplying and/or dividing them by well-chosen numbers.

5) Get the ratio in its simplest form, and that tells you the empirical formula of the compound.

EXAMPLE: Find the empirical formula of the iron oxide produced when 44.8 g of iron react with 19.2 g of oxygen. (A_r for iron = 56, A_r for oxygen = 16)

METHOD:

		Fe	O
1)	List the two elements:	**Fe**	**O**
2)	Write in the experimental masses:	44.8	19.2
3)	Divide by the A_r for each element:	$44.8/56 = 0.8$	$19.2/16 = 1.2$
4)	Multiply by 10...	8	12
	...then divide by 4:	2	3

5) So the simplest formula is 2 atoms of Fe to 3 atoms of O, i.e. Fe_2O_3.

This empirical method (i.e. based on experiment) is the only way of finding out the formula of a compound. Rust is iron oxide, sure, but is it FeO, or Fe_2O_3? Only an experiment to determine the empirical formula will tell you for certain.

One Mole of Gas Occupies a Volume of 24 dm³

Remember dm³ is just a fancy way of writing 'litre', so 1 dm³ = 1000 cm³

Remember moles from page 131? Well here's a wee bit more about them:

One mole of any gas always occupies 24 dm³ (= 24 000 cm³) at room temperature and pressure (RTP = 25 °C and 1 atmosphere)

Example 1: What's the volume of 4.5 moles of chlorine at RTP?

Answer: 1 mole = 24 dm³, so 4.5 moles = 4.5 × 24 dm³ = 108 dm³

Example 2: How many moles are there in 8280 cm³ of hydrogen gas at RTP?

Answer: Number of moles = $\dfrac{\text{Volume of gas}}{\text{Volume of 1 mole}}$ = $\dfrac{8.28}{24}$ = 0.345 moles   *Don't forget to convert from cm³ to dm³.*

It's all about making the ratio as simple as possible...

If you find these scary, just keep practising using the stepwise method until you've mastered it.

Conservation of Mass

When it comes to chemical reactions, there's one thing that you <u>really</u> need to know —
mass <u>stays the same</u> before, during and after a reaction.

In a Chemical Reaction, *Mass* is *Always Conserved*

1) During a chemical reaction <u>no atoms are destroyed</u> and <u>no atoms are created</u>.

2) This means there are the <u>same number and types of atoms</u> on each side of a reaction equation.

3) Because of this no mass is lost or gained — we say that mass is <u>conserved</u> during a reaction.

<u>Example</u>: $2Li + F_2 \rightarrow 2LiF$
<u>Method</u>: There are <u>2</u> lithium atoms and <u>2</u> fluorine atoms on <u>each side</u> of the equation.

4) By adding up the relative formula masses on each side of the equation
you can see that mass is conserved.

<u>Example</u>:	$2Li$	+	F_2	$\rightarrow$	$2LiF$
<u>Method</u>:	(2×7)	+	(2×19)	$\rightarrow$	2×26
	14	+	38	$\rightarrow$	52
		52		$\rightarrow$	52

So, <u>mass is conserved</u> in this equation.

5) You can use simple <u>ratios</u> to calculate the reacting masses in a reaction.

<u>Example</u>: In the reaction, $2Li + F_2 \rightarrow 2LiF$, 14 g of lithium will react with 38 g of fluorine.
<u>Method</u>: The only product that's formed is lithium fluoride,
so 14 + 38 = 52 g will be produced.
The masses for this reaction will always be in the same proportions as this.
Multiplying or dividing these masses by the same number gives you other
sets of reacting masses.

Element / compound in reaction	Lithium	Fluorine	Lithium fluoride
Original reacting masses	14 g	38 g	52 g
Reacting masses set 2	14 ÷ 2 = 7 g	38 ÷ 2 = 19 g	52 ÷ 2 = 26 g
Reacting masses set 3	14 × 1.5 = 21 g	38 × 1.5 = 57 g	52 × 1.5 = 78 g

Atoms — they're never created or destroyed in reactions...

... so that's why the mass always stays the same. If there are 57 g of reactants, you're going to get 57 g
of products. Or if there are 81 g of reactants, you'll get 81 g of products. You get the picture...

Calculating Masses in Reactions

You can also work out masses of reactants (starting materials) and products in reactions.

The Three Important Steps — *Not to Be Missed...*

1) <u>Write out</u> the balanced <u>equation</u>.

2) <u>Work out</u> M_r — just for the <u>two bits you want</u>.

3) Apply the rule: <u>Divide to get one, then multiply to get all</u>.
   (But you have to apply this first to the substance they
   give you information about, and then the other one!)

Don't worry — these steps should all make sense when you look at the example below.

Example:

What mass of magnesium oxide is produced when 60 g of magnesium is burned in air?

<u>Answer:</u>

1) Write out the <u>balanced equation</u>:

$$2Mg + O_2 \rightarrow 2MgO$$

2) Work out the <u>relative formula masses</u>:
   (don't do the oxygen — you don't need it)

$$2 \times 24 \rightarrow 2 \times (24 + 16)$$
$$48 \rightarrow 80$$

3) Apply the rule: <u>Divide to get one, then multiply to get all</u>:

The two numbers, 48 and 80, tell us that 48 g of Mg react to give 80 g of MgO. Here's the tricky bit. You've now got to be able to write this down:

> 48 g of Mg ...... reacts to give ...... 80g of MgO
> 1 g of Mg ...... reacts to give ......
> 60 g of Mg ...... reacts to give ......

<u>The big clue</u> is that in the question they've said we want to burn "<u>60 g of magnesium</u>", i.e. they've told us how much <u>magnesium</u> to have, and that's how you know to write down the <u>left-hand side</u> of it first, because:

**We'll first need to ÷ by 48 to get 1 g of Mg
and then need to × by 60 to get 60 g of Mg.**

<u>Then</u> you can work out the numbers on the other side (shown in blue below) by realising that you must <u>divide both sides by 48</u> and then <u>multiply both sides by 60</u>.

÷ 48
× 60

48 g of Mg .... 80 g of MgO
1 g of Mg .... 1.67 g of MgO
60 g of Mg .... 100 g of MgO

÷ 48
× 60

The mass of product is called the yield of a reaction. You should realise that in practice you never get 100% of the yield, so the amount of product will be slightly less than calculated (see page 140).

This finally tells us that <u>60 g of magnesium will produce 100 g of magnesium oxide</u>.

If the question had said "Find how much magnesium gives 500 g of magnesium oxide", you'd fill in the MgO side first, <u>because that's the one you'd have the information about</u>.

Warm-Up and Exam Questions

Lots to remember on the last few pages. Try these and see how good your understanding really is.

Warm-Up Questions

1) What name is given to the sum of the relative atomic masses of the atoms in a molecule?
2) Write down the definition of a mole.
3) What is the mass of one mole of oxygen gas?
4) How many moles of fluorine gas are in 2.4 dm³ at RTP?

Exam Questions

1 Use the A_r values B = 11, O = 16, F = 19 and H = 1 to calculate the relative formula masses of these boron compounds:

   (a) BF_3

   (1 mark)

   (b) $B(OH)_3$

   (1 mark)

2 Analysis of a gaseous oxide of sulfur shows that it contains 60% oxygen by mass. (A_r values: S = 32, O = 16.)

   (a) What is the percentage mass of sulfur in the oxide?

   (1 mark)

   (b) Work out the formula of the oxide.

   (2 marks)

   (c) Calculate how many moles there are in 2460 cm³ of the oxide at RTP.

   (2 marks)

3 Heating a test tube containing 2 g of calcium carbonate produced 1.08 g of calcium oxide when it was reweighed. The equation for the reaction is:

$$CaCO_3(s) \rightarrow CaO(s) + CO_2(g)$$

(M_r values: $CaCO_3$ = 100, CaO = 56.)

   (a) Calculate the amount of calcium oxide you would expect to be formed from 2 g of calcium carbonate.

   (1 mark)

   (b) Compare the value to the mass obtained in the experiment. Suggest a possible reason for the difference.

   (1 mark)

Exam Questions

4   A good fertiliser contains lots of nitrogen for plant growth. An agricultural scientist needs to compare the amount of nitrogen in each of the three different fertilisers listed below.

Fertiliser	Formula
urea	$CO(NH_2)_2$
potassium nitrate	KNO_3
ammonium nitrate	NH_4NO_3

(A_r values: C = 12, O = 16, N = 14, H = 1, K = 39.)

(a) Work out the percentage mass of nitrogen in each of the three fertilisers.

(6 marks)

(b) Using your answers to part (a), explain which one of the three would you expect to make the best fertiliser.

(2 marks)

5   (a) Calculate the relative formula mass of sodium hydroxide, NaOH.
(A_r values: Na = 23, O = 16, H = 1.)

(1 mark)

(b) How many moles are there in 4 g of sodium hydroxide?

(2 marks)

(c) In the following reaction, 246 g of reactants produce 6 g of hydrogen.

sodium + water → sodium hydroxide + hydrogen

What mass of sodium hydroxide, NaOH, will be produced? Explain your answer.

(2 marks)

6   A sample of an organic hydrocarbon was burnt completely in air.
4.4 g of carbon dioxide and 1.8 g of water were formed.

(a) Calculate the number of moles of carbon in the hydrocarbon.
(Relative atomic masses: H = 1, C = 12, O = 16.)

(2 marks)

(b) Calculate the number of moles of hydrogen in the hydrocarbon.

(2 marks)

(c) Use your answers to parts (a) and (b) to work out the empirical formula of the hydrocarbon.

(1 mark)

7   60 g of calcium react with 106.5 g of chlorine to produce calcium chloride.
Calculate the empirical formula of calcium chloride. Show all your working.
(Relative atomic masses: Ca = 40, Cl = 35.5.)

(2 marks)

Atom Economy

It's important in industrial reactions that as much of the reactants as possible get turned into useful products. This depends on the <u>atom economy</u> and the <u>percentage yield</u> (see page 140) of the reaction.

"Atom Economy" — % of Reactants Changed to Useful Products

1) A lot of reactions make <u>more than one product</u>. Some of them will be <u>useful</u>, but others will just be <u>waste</u>, e.g. when you make calcium oxide from limestone, you also get CO_2 as a waste product.

2) The <u>atom economy</u> of a reaction tells you how much of the <u>mass</u> of the reactants is wasted when manufacturing a chemical.

    Here's the equation:

$$\text{atom economy} = \frac{\text{total } M_r \text{ of desired products}}{\text{total } M_r \text{ of all products}} \times 100$$

3) <u>100%</u> atom economy means that <u>all</u> the atoms in the reactants have been turned into <u>useful</u> (desired) <u>products</u>. The <u>higher</u> the atom economy the 'greener' the process.

Example

Hydrogen gas is made on a large scale by reacting natural gas (methane) with steam.

$$CH_4(g) + H_2O(g) \rightarrow CO(g) + 3H_2(g)$$

Calculate the atom economy of this reaction.

<u>Method</u>:

1) <u>Identify</u> the useful product — that's the <u>hydrogen gas</u>.

2) Work out the M_r of <u>all the products</u> and the <u>useful product</u>:

CO	3H₂	3H₂
12 + 16	3 × (2 × 1)	3 × (2 × 1)
34		6

3) Use the <u>formula</u> to calculate the atom economy:

$$\text{atom economy} = \frac{6}{34} \times 100 = \underline{17.6\%}$$

So in this reaction, <u>over 80%</u> of the starting materials are <u>wasted</u>.

Atom Economy

As you may have guessed, a <u>high</u> atom economy is <u>good</u>, and a <u>low</u> atom economy is <u>not so good</u>.

High Atom Economy is Better for **Profits** and the **Environment**

1) Pretty obviously, if you're making <u>lots of waste</u>, that's a <u>problem</u>.

2) Reactions with low atom economy <u>use up resources</u> very quickly.

3) At the same time, they produce loads of <u>waste</u> materials that have to be <u>disposed</u> of somehow.

4) That tends to make these reactions <u>unsustainable</u> — the raw materials will run out and the waste has to go somewhere.

5) For the same reasons, low atom economy reactions aren't usually <u>profitable</u>.

6) Raw materials are <u>expensive to buy</u>, and waste products can be expensive to <u>remove</u> and dispose of <u>responsibly</u>.

7) The best way around the problem is to find a <u>use</u> for the waste products rather than just <u>throwing them away</u>.

8) There's often <u>more than one way</u> to make the product you want, so the trick is to come up with a reaction that gives <u>useful</u> <u>"by-products"</u> rather than useless ones.

9) The reactions with the <u>highest</u> atom economy are the ones that only have <u>one product</u>.

10) Those reactions have an atom economy of <u>100%</u>.

Atom economy — important, but not the whole story...

Atom economy isn't the only thing that affects profits — there are other costs besides buying raw materials and disposing of waste. There are <u>energy</u> and <u>equipment</u> costs, as well as the cost of <u>paying people</u> to work at the plant. You need to think about the <u>percentage yield</u> of the reaction too (p.140).

Percentage Yield

Percentage yield tells you about the <u>overall success</u> of an experiment. It compares what you calculate you should get (<u>predicted yield</u>) with what you get in practice (<u>actual yield</u>).

Percentage Yield Compares *Actual* and *Predicted* Yield

The amount of product you get is known as the <u>yield</u>. The more reactants you start with, the higher the <u>actual yield</u> will be — that's pretty obvious. But the <u>percentage yield doesn't</u> depend on the amount of reactants you started with — it's a <u>percentage</u>.

1) The <u>predicted yield</u> of a reaction can be calculated from the <u>balanced reaction equation</u> (see page 23).

2) Percentage yield is given by the formula:

The predicted yield is sometimes called the theoretical yield.

$$\text{percentage yield} = \frac{\text{actual yield (grams)}}{\text{predicted yield (grams)}} \times 100$$

3) Percentage yield is <u>always</u> somewhere between 0 and 100%.

4) A 100% percentage yield means that you got <u>all</u> the product you expected to get.

5) A 0% yield means that <u>no</u> reactants were converted into product, i.e. no product at all was <u>made</u>.

Yields are Always **Less than 100%**

Even though <u>no atoms are gained or lost</u> in reactions, in real life, you <u>never</u> get a 100% percentage yield. Some product or reactant <u>always</u> gets lost along the way — and that goes for big <u>industrial processes</u> as well as school lab experiments.

Lots of things can go wrong, but you can find five of these things conveniently located on the next page.

Even with the best equipment, you can't get the maximum product

A high percentage yield means there's <u>not much waste</u> — which is good for <u>preserving resources</u> and keeping production <u>costs down</u>. If a reaction's going to be worth doing commercially, it generally has to have a high percentage yield or recyclable reactants. There's a <u>formula</u> for working out the all important percentage yield — you just need to know what you got and what you expected to get.

Percentage Yield and Reversible Reactions

There are **Several Reasons** Why Yields **Can't** be 100%

1) The reaction is **reversible**

> A <u>reversible reaction</u> is one where the <u>products</u> of the reaction can <u>themselves react</u> to produce the <u>original reactants</u>
>
> $$A + B \rightleftharpoons C + D$$
>
> <u>For example:</u>
> ammonium chloride $\rightleftharpoons$ ammonia + hydrogen chloride

This means that the reactants will never be completely converted to products because the reaction goes both ways. Some of the <u>products</u> are always <u>reacting together</u> to change back to the original reactants. This will mean a <u>lower yield</u>.

2) **Filtration**

1) When you <u>filter a liquid</u> to remove <u>solid particles</u>, you nearly always <u>lose</u> a bit of liquid or a bit of solid.
2) So, some of the product may be lost when it's <u>separated</u> from the reaction mixture.

3) **Transferring liquids**

You always lose a bit of liquid when you <u>transfer</u> it from one container to another — even if you manage not to spill it. Some of it always gets left behind on the <u>inside surface</u> of the old container.

4) **Evaporation**

Liquids evaporate <u>all the time</u> — and even more so while they're being heated.

5) **Unexpected reactions**

1) Things don't always go exactly to plan. Sometimes there can be other <u>unexpected reactions</u> happening which <u>use up the reactants</u>.
2) This means there's not as much reactant to make the <u>product</u> you want.

Product Yield is Important for **Sustainable Development**

1) Thinking about product yield is important for <u>sustainable development</u>.

2) Sustainable development is about making sure that we don't use <u>resources</u> faster than they can be <u>replaced</u> — there needs to be enough for <u>future generations</u> too.

3) So, for example, using as <u>little energy</u> as possible to create the <u>highest product yield possible</u> means that resources are <u>saved</u>. A low yield means wasted chemicals — not very sustainable.

Warm-Up and Exam Questions

Time to see what you can remember about atom economy and percentage yield...

Warm-Up Questions

1) What effect does a waste by-product have on the atom economy of a reaction?
2) What is the atom economy of the reaction shown? $2SO_2 + O_2 \rightarrow 2SO_3$
3) Why might a reaction with a low atom economy be bad for the environment?
4) What is the percentage yield of a reaction which produced 4 g of product if the predicted yield was 5 g?
5) Why might a reaction with a low percentage yield be bad for sustainable development?

Exam Questions

1   Ethanol produced by the fermentation of sugar can be converted into ethene, as shown below. The ethene can then be used to make polythene.

$$C_2H_6O \text{ (g)} \quad \rightarrow \quad C_2H_4 \text{ (g)} + H_2O \text{ (g)}$$

Calculate the atom economy of this reaction. (A_r values: C = 12, O = 16, H = 1.)

(3 marks)

2   A sample of copper was made by reducing 4 g of copper oxide with methane gas. When the black copper oxide turned orange-red, the sample was scraped out into a beaker. Sulfuric acid was added to dissolve any copper oxide that remained. The sample was then washed, filtered and dried. 2.8 g of copper was obtained. (A_r values: Cu = 63.5, O = 16.)

The equation for this reaction is: $CH_4 + 4CuO \rightarrow 4Cu + 2H_2O + CO_2$

(a)   Use the equation to calculate the maximum mass of copper which could be obtained from the reaction (the predicted yield).

(3 marks)

(b)   Calculate the percentage yield of the reaction.

(2 marks)

(c)   Suggest three different reasons why the yield of the reaction was less than 100%.

(3 marks)

3   *In this question you will be assessed on the quality of your English, the organisation of your ideas and your use of appropriate specialist vocabulary.*

Discuss the reasons why yields from chemical reactions are always less than 100%.

(6 marks)

4   Which of the following statements about atom economy is **not** true?

A   Reactions that only have one product have a very high atom economy.
B   Reactions with a low atom economy are unsustainable.
C   Reactions with a low atom economy are not usually profitable.
D   Reactions with a high atom economy use up resources quickly.

(1 mark)

Revision Summary for Section 7

Some more tricky questions to stress you out. The thing is though, why bother doing easy questions? These meaty monsters find out what you really know, and worse, what you really don't. Yeah, I know, some of them are a bit scary, but if you want to get anywhere in life you've got to face up to a bit of hardship. That's just the way it is. Take a few deep breaths and then try these.

1) Define relative formula mass.

2)* Find A_r or M_r for these (use the periodic table at the front of the book):
   a) Ca   b) Ag   c) CO_2 d) $MgCO_3$ e) Na_2CO_3 f) ZnO   g) KOH   h) NH_3

3) What is the link between moles and relative formula mass?

4)* a) Calculate the percentage mass of carbon in:
       i) $CaCO_3$ ii) CO_2 iii) CH_4
   b) Calculate the percentage mass of metal in:
       i) Na_2O ii) Fe_2O_3 iii) Al_2O_3

5) a) What is an empirical formula?
   b)* Find the empirical formula of the compound formed when
       21.9 g of magnesium, 29.2 g of sulfur and 58.4 g of oxygen react.

6) What volume does one mole of gas take up at room temperature and pressure?

7)* What mass of sodium is needed to produce 108.2 g of sodium oxide (Na_2O)?

8) Write the equation for calculating the atom economy of a reaction.

9) Explain why it is important to use industrial reactions with a high atom economy.

10) Describe three factors that can reduce the percentage yield of a reaction.

* Answers on page 258

Hazard and State Symbols

Chemistry is full of useful symbols — there are hazard symbols for warning you about chemicals that could potentially cause harm, and state symbols for telling you something's physical state.

Hazard Symbols Are There To Warn You About Danger

Lots of the chemicals you'll meet in Chemistry can be bad for you or dangerous in some way. That's why the chemical containers will normally have symbols on them to tell you what the dangers are. Understanding these hazard symbols means that you'll be able to use suitable safe-working procedures in the lab.

Highly Flammable
Catches fire easily.
Example: Petrol.

Harmful
Like toxic but not quite as dangerous.
Example: Copper sulfate.

Explosive
Can explode – BANG.
Example: Some peroxides.

Oxidising
Provides oxygen which allows other materials to burn more fiercely.
Example: Liquid oxygen.

Irritant
Not corrosive but can cause reddening or blistering of the skin.
Examples: Bleach, children, etc.

Corrosive
Attacks and destroys living tissues, including eyes and skin.
Example: Concentrated sulfuric acid.

Toxic
Can cause death either by swallowing, breathing in, or absorption through the skin.
Example: Hydrogen cyanide.

State Symbols Tell You What Physical State It's In

These are easy enough, so make sure you know them — especially aq (aqueous).

(s) — Solid	(l) — Liquid	(g) — Gas	(aq) — Dissolved in water

E.g. $2Mg_{(s)} + O_{2\,(g)} \rightarrow 2MgO_{(s)}$

Acids and Alkalis

Testing the pH of a solution means using an <u>indicator</u> — and that means pretty <u>colours</u>...

The pH scale Goes From 0 to 14

1) The <u>pH scale</u> is a measure of how <u>acidic</u> or <u>alkaline</u> a solution is.
2) The <u>strongest acid</u> has <u>pH 0</u>. The <u>strongest alkali</u> has <u>pH 14</u>.
3) A <u>neutral</u> substance has <u>pH 7</u> (e.g. pure water).

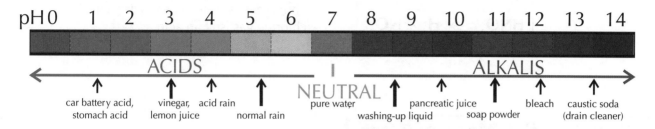

pH 0 1 2 3 4 5 6 7 8 9 10 11 12 13 14

ACIDS | ALKALIS
NEUTRAL

car battery acid, stomach acid   vinegar, lemon juice   acid rain   normal rain   pure water   washing-up liquid   pancreatic juice   soap powder   bleach   caustic soda (drain cleaner)

An Indicator is Just a Dye That Changes Colour

1) The dye in an indicator <u>changes colour</u> depending on whether it's <u>above or below a certain pH</u>.
2) <u>Universal indicator</u> is a <u>combination of dyes</u> which gives the colours shown above.
3) It's very useful for <u>estimating</u> the pH of a solution.

Acids and Bases Neutralise Each Other

An <u>ACID</u> is a substance with a pH of less than 7. Acids form <u>H^+ ions</u> in <u>water</u>.
A <u>BASE</u> is a substance with a pH of greater than 7.
An <u>ALKALI</u> is a base that <u>dissolves in water</u>. Alkalis form <u>OH^- ions</u> in <u>water</u>.
So, <u>H^+</u> ions make solutions <u>acidic</u> and <u>OH^-</u> ions make them <u>alkaline</u>.

The reaction between acids and bases is called <u>neutralisation</u>:

$$acid \ + \ base \ \rightarrow \ salt \ + \ water$$

Neutralisation can also be seen in terms of <u>H^+</u> and <u>OH^- ions</u> like this:

$$H^+_{(aq)} \ + \ OH^-_{(aq)} \ \rightarrow \ H_2O_{(l)}$$

Hydrogen (H^+) ions react with hydroxide (OH^-) ions to produce water.

When an acid neutralises a base (or vice versa), the <u>products</u> are <u>neutral</u>, i.e. they have a <u>pH of 7</u>.
An indicator can be used to show that a neutralisation reaction is over (Universal indicator will go green).

Interesting(ish) fact — your skin is slightly acidic (pH 5.5)...

It might sound like a bit of a complicated chemical idea, but neutralisation is just a hydrogen ion reacting with a hydroxide ion to make water. That's it. Not so complicated after all really...

Strong and Weak Acids

Right then. Strong acids versus weak acids. Brace yourself.

Acids **Produce Protons** in **Water**

The thing about acids is that they ionise — they produce hydrogen ions, H^+. *An H^+ ion is just a proton.*
For example,

$$HCl \rightarrow H^+ + Cl^-$$
$$HNO_3 \rightarrow H^+ + NO_3^-$$

But HCl doesn't produce hydrogen ions until it meets water — so hydrogen chloride gas isn't an acid.

Acids Can Be **Strong** or **Weak**

1) Strong acids (e.g. sulfuric, hydrochloric and nitric) ionise completely in water. This means every hydrogen atom releases a hydrogen ion — so there are loads of H^+ ions.

2) Weak acids (e.g. ethanoic, citric, carbonic) do not fully ionise. Only some of the hydrogen atoms in the compound release hydrogen ions — so only small numbers of H^+ ions are formed.

For example,

Strong acid: $HCl \rightarrow H^+ + Cl^-$

Weak acid: $CH_3COOH \rightleftharpoons H^+ + CH_3COO^-$

Use a 'reversible reaction' arrow for a weak acid.

3) The ionisation of a weak acid is a reversible reaction which sets up an equilibrium mixture. Since only a few H^+ ions are released, the equilibrium lies well to the left.

4) The pH of an acid or alkali is a measure of the concentration of H^+ ions in the solution. Strong acids typically have a pH of about 1 or 2, while the pH of a weak acid might be 4, 5 or 6.

5) The pH of an acid or alkali can be measured with a pH meter or with universal indicator paper (or can be estimated by seeing how fast a sample reacts with, say, magnesium).

Weak acids don't fully ionise, strong acids ionise completely

Acids are acidic because of H^+ ions. And strong acids are strong because they let go of all their H^+ ions at the drop of a hat... well, at the drop of a drop of water. This is tricky — no doubt about it, but if you can get your head round this, then you can probably cope with just about anything.

Strong and Weak Acids

Strong acids and weak acids sometimes <u>behave differently</u>.

Don't Confuse **Strong** Acids With **Concentrated** Acids

1) Acid <u>strength</u> (i.e. strong or weak) tells you <u>what proportion</u> of the acid molecules <u>ionise</u> in water.

2) The <u>concentration</u> of an acid is different. Concentration measures <u>how many moles of acid molecules</u> there are in a litre (1 dm^3) of water. Concentration is basically how <u>watered down</u> your acid is.

3) Note that concentration describes the <u>total number</u> of dissolved acid molecules — <u>not</u> the number of molecules that produce hydrogen ions.

4) The more moles of acid molecules per dm^3, the <u>more concentrated</u> the acid is.

5) So you can have a <u>dilute but strong</u> acid, or a <u>concentrated but weak</u> acid.

Strong Acids are **Better Electrical Conductors** Than Weak Acids

1) Ethanoic acid has a <u>much lower electrical conductivity</u> than the <u>same concentration</u> of hydrochloric acid. It's all to do with the concentration of the ions.

2) It's the ions that carry the charge through the acid solutions as they move. So the <u>lower concentration of ions</u> in the weak acid means <u>less charge</u> can be carried. Simple.

3) <u>Electrolysis</u> of hydrochloric acid or ethanoic acid <u>produces H$_2$</u> because they both produce H$^+$ ions.

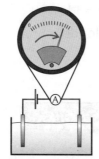

hydrochloric acid
(1 mol/dm^3)

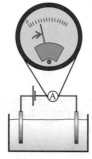

ethanoic acid
(1 mol/dm^3)

Strong isn't the same as concentrated...

Dilute, strong acids or concentrated, weak ones... this can all get a bit confusing. Just remember the concentration is how watered down your acid is, whereas strength is to do with the proportion of the acid molecules that ionise in water (see page 146). Confusing yes, but stick at it and you'll get there.

Acids Reacting With Metals

There are loads of different salts out there. Some of them are made when an <u>acid</u> reacts with a <u>metal</u>.

Metals react with Acids to give Salts

$$\text{acid} + \text{metal} \;\rightarrow\; \text{salt} + \text{hydrogen}$$

That's written big because it's really worth remembering. Here's the <u>typical experiment</u>:

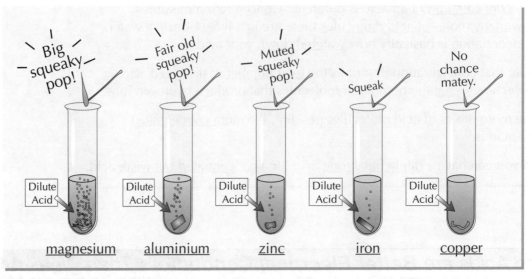

1) The more <u>reactive</u> the metal, the <u>faster</u> the reaction will go — very reactive metals (e.g. sodium) react <u>explosively</u>.

2) <u>Copper</u> does <u>not</u> react with dilute acids <u>at all</u> — because it's <u>less</u> reactive than <u>hydrogen</u>.

3) The <u>speed</u> of reaction is indicated by the <u>rate</u> at which the <u>bubbles</u> of hydrogen are given off.

4) The <u>hydrogen</u> is confirmed by the <u>burning splint test</u> giving the notorious '<u>squeaky pop</u>'.

5) The <u>name</u> of the <u>salt</u> produced depends on which <u>metal</u> is used, and which <u>acid</u> is used:

Hydrochloric Acid Will Always Produce Chloride Salts:

$2HCl + Mg \rightarrow MgCl_2 + H_2$ (Magnesium chloride)
$6HCl + 2Al \rightarrow 2AlCl_3 + 3H_2$ (Aluminium chloride)
$2HCl + Zn \rightarrow ZnCl_2 + H_2$ (Zinc chloride)

Sulfuric Acid Will Always Produce Sulfate Salts:

$H_2SO_4 + Mg \rightarrow MgSO_4 + H_2$ (Magnesium sulfate)
$3H_2SO_4 + 2Al \rightarrow Al_2(SO_4)_3 + 3H_2$ (Aluminium sulfate)
$H_2SO_4 + Zn \rightarrow ZnSO_4 + H_2$ (Zinc sulfate)

Nitric Acid Produces Nitrate Salts When NEUTRALISED, But...

Nitric acid reacts fine with alkalis, to produce nitrates, but it can play silly devils with metals and produce nitrogen oxides instead, so we'll ignore it here.

Oxides, Hydroxides and Carbonates

I'm afraid there's more stuff on <u>neutralisation</u> reactions coming up...

Metal *Oxides* and Metal *Hydroxides* Are *Bases*

1) Some <u>metal oxides</u> and <u>metal hydroxides</u> dissolve in <u>water</u>. These soluble compounds are <u>alkalis</u>.
2) Even bases that won't dissolve in water will still react with acids.
3) So, all <u>metal oxides</u> and <u>metal hydroxides</u> react with <u>acids</u> to form a <u>salt</u> and <u>water</u>.

acid + metal oxide → salt + water

acid + metal hydroxide → salt + water

These are neutralisation reactions.

The *Combination* of Metal and Acid Decides the *Salt*

This isn't exactly exciting but it's pretty easy, so try and get the hang of it:

hydrochloric acid + copper oxide → copper chloride + water
hydrochloric acid + sodium hydroxide → sodium chloride + water

sulfuric acid + zinc oxide → zinc sulfate + water
sulfuric acid + calcium hydroxide → calcium sulfate + water

nitric acid + magnesium oxide → magnesium nitrate + water
nitric acid + potassium hydroxide → potassium nitrate + water

The symbol equations are all pretty much the same. Here are two of them:

$$H_2SO_{4\ (aq)} + ZnO_{(s)} \rightarrow ZnSO_{4\ (aq)} + H_2O_{(l)}$$
$$HNO_{3\ (aq)} + KOH_{(aq)} \rightarrow KNO_{3\ (aq)} + H_2O_{(l)}$$

Metal *Carbonates* Give *Salt + Water + Carbon Dioxide*

More gripping reactions involving acids. At least there are some <u>bubbles</u> involved here.

acid + metal carbonate → salt + water + carbon dioxide

The reaction is the same as any other neutralisation reaction EXCEPT that <u>carbonates</u> give off <u>carbon dioxide</u> as well. Here's an example for you:

hydrochloric acid + sodium carbonate → sodium chloride + water + carbon dioxide

$$2HCl_{(aq)} + Na_2CO_{3(s)} \rightarrow 2NaCl_{(aq)} + H_2O_{(l)} + CO_{2(g)}$$

Warm-Up and Exam Questions

Well, that's hazard symbols and acids and alkalis done — so it must be time for some questions.
There are a few warm-up questions to ease you in before you get stuck into the exam questions.

Warm-Up Questions

1) What is the meaning of the hazard symbol shown on the right?
2) What name is given to the type of reaction in which an acid reacts with a base?
3) What's the difference between a strong acid and a weak acid?
4) Which two substances are formed when nitric acid reacts with copper oxide?
5) Which three substances are formed when sulfuric acid reacts with calcium carbonate?

Exam Questions

1   The table shows the results when five solutions, A–E, were tested with Universal indicator.

Solution	Colour	pH
A		1
B	pale green	
C	orange	5
D	dark blue	
E		14

(a) Complete the blanks in the table.

(2 marks)

(b) Which solution is a weak acid?

(1 mark)

(c) Which solution is a strong alkali?

(1 mark)

(d) Which solution contains sodium chloride?

(1 mark)

(e) Which solution is battery acid?

(1 mark)

2   All metal oxides and metal hydroxides react with acids to form a salt and water.

(a) What are metal oxides and metal hydroxides that dissolve in water called?

(1 mark)

(b) (i)   Complete the symbol equation given below for the reaction of a metal oxide with an acid.

$MgO + 2HCl \rightarrow$ _____ + _____

(2 marks)

     (ii) Write out the word equation for the above reaction.

(1 mark)

(c) Acids also react with carbonates to form a salt and water.
     What other substance is also produced?

(1 mark)

(d) Complete and balance the symbol equation given below for the reaction of nitric acid with copper carbonate.

_____ $+ CuCO_3 \rightarrow Cu(NO_3)_2 +$ _____ $+$ _____

(2 marks)

Exam Questions

3   A student conducts a study looking at how different metals react with acid.
He places small pieces of different metals into test tubes containing dilute hydrochloric acid.
He then records what he sees in a table.
The table below shows the student's results.

Metal	Bubbles
Magnesium	Loads
Aluminium	A lot
Zinc	Some
Copper	None

(a)   Give the general word equation for reacting an acid with a metal.

(1 mark)

(b)   Explain how you can tell from the results that magnesium is more reactive
than copper.

(3 marks)

(c)   The student holds a burning splint above the test tube containing zinc and
hydrochloric acid.
(i)   Describe what will happen.

(1 mark)

(ii)   The student holds a burning splint over the test tube containing copper
and acid. Suggest whether or not the result of the test will be different.
Explain your answer.

(2 marks)

(d)   The student repeats the experiment again, this time using sulfuric acid.
Write out a balanced symbol equation to show the reaction between
magnesium and sulfuric acid.

(1 mark)

4   A student has two acids, acid A and acid B. She wants to find out which acid is stronger so
decides to measure the electrical conductivity of the acid using the apparatus below.

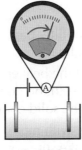
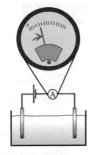

Acid A   **Acid B**

From the meter readings she can tell which acid is stronger.
Which is the strongest acid? Explain your answer.

(5 marks)

Making Insoluble Salts

Some salts are <u>soluble</u> and some are <u>insoluble</u> — it's just the way the cookie crumbles.

The **Rules** of **Solubility**

This table is a pretty fail-safe way of working out whether a substance is soluble in water or not.

Substance	Soluble or Insoluble?
common salts of sodium, potassium and ammonium	soluble
nitrates	soluble
common chlorides	soluble (except silver chloride and lead chloride)
common sulfates	soluble (except lead, barium and calcium sulfate)
common carbonates and hydroxides	insoluble (except for sodium, potassium and ammonium ones)

Making **Insoluble** Salts — **Precipitation** Reactions

1) To make a pure, dry sample of an <u>insoluble</u> salt, you can use a <u>precipitation reaction</u>. You just need to pick the right two <u>soluble salts</u>, they <u>react</u> and you get your <u>insoluble salt</u>.

2) E.g. to make <u>lead chloride</u> ⟹ (insoluble), mix <u>lead nitrate</u> and <u>sodium chloride</u> (both soluble).

> lead nitrate + sodium chloride → lead chloride + sodium nitrate
> $Pb(NO_3)_{2\,(aq)} + 2NaCl_{(aq)} \longrightarrow PbCl_{2\,(s)} + 2NaNO_{3\,(aq)}$

Method

STAGE 1

1) Add 1 spatula of <u>lead nitrate</u> to a test tube, and fill it with <u>distilled water</u>. <u>Shake it thoroughly</u> to ensure that all the lead nitrate has <u>dissolved</u>. Then do the same with 1 spatula of <u>sodium chloride</u>. (Use distilled water to make sure there are <u>no other ions</u> about.)

2) Tip the <u>two solutions</u> into a small beaker, and give it a good stir to make sure it's all mixed together. The lead chloride should <u>precipitate</u> out.

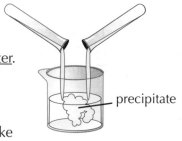

precipitate

STAGE 2

filter paper

filter funnel

1) Put a folded piece of <u>filter paper</u> into a <u>filter funnel</u>, and stick the funnel into a <u>conical flask</u>.

2) <u>Pour</u> the contents of the beaker into the middle of the filter paper. (Make sure that the solution doesn't go above the filter paper — otherwise some of the solid could dribble down the side.)

3) <u>Swill out</u> the beaker with more distilled water, and tip this into the filter paper — to make sure you get <u>all the product</u> from the beaker.

STAGE 3

1) Rinse the contents of the filter paper with distilled water to make sure that <u>all the soluble sodium nitrate</u> has been washed away.

2) Then just scrape the <u>lead chloride</u> onto fresh filter paper and leave to dry.

lead chloride

Making Soluble Salts

When an acid and an alkali react you get a <u>salt</u>. However, if the salt that's made is <u>soluble</u>, getting hold of it can be tricky. That's where a bit of cunning comes in...

Making **Soluble Salts** Using an **Acid** and an **Insoluble Reactant**

1) You can make soluble salts by reacting an acid with an insoluble base.

2) You need to pick the right <u>acid</u>, plus a <u>metal</u> or an <u>insoluble base</u> (a <u>metal oxide</u> or <u>metal hydroxide</u>).

For example, you can add <u>copper oxide</u> to <u>hydrochloric acid</u> to make <u>copper chloride</u>:

$$CuO_{(s)} + 2HCl_{(aq)} \rightarrow CuCl_{2\,(aq)} + H_2O_{(l)}$$

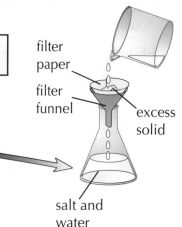

filter paper

filter funnel

excess solid

3) You add the <u>base</u> to the <u>acid</u> — the solid will <u>dissolve</u> in the acid as it reacts. You will know when all the acid has been neutralised because the excess solid will just <u>sink</u> to the bottom of the flask.

4) Then <u>filter</u> out the <u>excess</u> solid to get a solution containing only salt and water.

5) Finally, <u>heat it gently</u> to slowly <u>evaporate</u> off the water and crystallise the salt.

salt and water

Making **Soluble Salts** Using an **Acid** and a **Soluble Reactant**

1) Soluble salts can also be made by reacting an acid with an <u>alkali</u> (soluble base) like <u>sodium</u>, <u>potassium</u> or <u>ammonium hydroxides</u>.

2) But you can't tell whether the reaction has <u>finished</u> — there's no signal that all the acid has been neutralised. You also can't just add an <u>excess</u> of alkali to the acid and filter out what's left because the salt is <u>soluble</u>.

3) Instead, you have to add <u>exactly</u> the right amount of alkali to <u>neutralise</u> the acid. So, you must first carry out a <u>titration</u> (see page 194) to work out the <u>exact amount</u> of alkali needed.

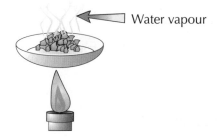

Water vapour

4) Then, carry out the reaction using exactly the right proportions of alkali and acid. You <u>won't need</u> to use an <u>indicator</u> this time because you know the volumes needed. So, the salt <u>won't be contaminated</u> with indicator.

5) The <u>solution</u> that remains when the reaction is complete contains only the <u>salt</u> and <u>water</u>. <u>Evaporate</u> off the water slowly and you'll be left with a <u>pure</u>, <u>dry</u> salt.

For example, <u>sulfuric acid</u> can be reacted with <u>sodium hydroxide</u> to make <u>sodium sulfate</u>:

$$H_2SO_{4(aq)} + 2NaOH_{(aq)} \rightarrow Na_2SO_{4(aq)} + 2H_2O_{(l)}$$

Get two solutions, mix 'em together — job's a good'un...

If you're making a soluble salt, you need to think <u>carefully</u> about what <u>chemicals</u> you'd need to get the salt you want and what method you should use. Whether you're making a soluble salt or an insoluble salt, you'll be left with just <u>salt</u> and <u>water</u> at the end. All you need to do is separate them.

Redox Reactions

In chemistry, things get oxidised and reduced all the time.

If *Electrons are Transferred*, *It's a Redox Reaction*

1) Oxidation can mean the <u>addition of oxygen</u> (or a reaction with it), and reduction can be the <u>removal of oxygen</u>, but on this page we're looking at oxidation and reduction in terms of <u>electrons</u>.

2) A <u>loss of electrons</u> is called <u>oxidation</u>.
   A <u>gain of electrons</u> is called <u>reduction</u>.

3) REDuction and OXidation happen <u>at the same time</u>
   — hence the term "REDOX".

4) An <u>oxidising agent</u> accepts electrons and <u>gets reduced</u>.

5) A <u>reducing agent</u> donates electrons and <u>gets oxidised</u>.

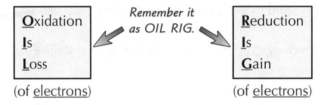

Oxidation **I**s **L**oss   *Remember it as OIL RIG.*   **R**eduction **I**s **G**ain

(of <u>electrons</u>)   (of <u>electrons</u>)

Some *Examples* of Redox Reactions:

Chlorine gas and iron(II) salt

1) <u>Chlorine gas</u> is passed into a solution of an iron(II) salt. The solution turns from green to yellow as the iron(II) ion is oxidised to iron(III). The Fe^{2+} ion <u>loses an electron</u> to form Fe^{3+}.
   $$Fe^{2+} - e^- \rightarrow Fe^{3+}$$

2) The <u>chlorine</u> causes this to happen — it's the <u>oxidising agent</u>.

3) The chlorine must've <u>gained</u> the electron that the Fe^{2+} lost. The chlorine's been <u>reduced</u>. The iron(II) ion must be the <u>reducing agent</u>.
   $$\tfrac{1}{2}Cl_2 + e^- \rightarrow Cl^-$$

Iron and dilute acid

1) <u>Iron atoms</u> are <u>oxidised</u> to iron(II) ions when they react with <u>dilute acid</u>.

2) The <u>iron atoms lose electrons</u>.
   They're <u>oxidised</u> by the hydrogen ions.
   $$Fe - 2e^- \rightarrow Fe^{2+}$$

3) The <u>hydrogen ions gain electrons</u>.
   They're <u>reduced</u> by the iron atoms.
   $$2H^+ + 2e^- \rightarrow H_2$$

Remember OIL RIG — 'Oxidation Is Loss, Reduction Is Gain'...

...it's a pretty handy way of reminding yourself about what goes on during a <u>redox reaction</u>. Make sure that you don't forget that redox reactions are all about the <u>transfer of electrons</u>. Got it. Good.

Redox Reactions

Can't get enough of redox reactions... it's your lucky day, there are some more coming your way.

Displacement Reactions are Redox Reactions

1) Displacement reactions involve one metal kicking another one out of a compound. Here's the rule:

> A MORE REACTIVE metal will displace a LESS REACTIVE metal from its compound.

2) If you put a reactive metal into the solution of a dissolved metal compound, the reactive metal will replace the less reactive metal in the compound.

Example: Put iron in a solution of tin(II) sulfate and the more reactive iron will "kick out" the less reactive tin from the solution. You end up with iron(II) sulfate solution and tin metal.

MAGNESIUM	Mg	
ZINC	Zn	reactivity ↑
IRON	Fe	
TIN	Sn	

$$\text{iron} + \text{tin(II) sulfate} \rightarrow \text{iron(II) sulfate} + \text{tin}$$
$$Fe_{(s)} + SnSO_{4(aq)} \rightarrow FeSO_{4(aq)} + Sn_{(s)}$$

In this reaction the iron loses 2 electrons to become a 2^+ ion — it's oxidised.
The tin ion gains these 2 electrons to become a tin atom — it's reduced.

$$Fe + SO_4^{2-} \rightarrow FeSO_4 + 2e^-$$

$$Sn^{2+} + 2e^- \rightarrow Sn$$

3) In displacement reactions it's always the metal ion that gains electrons and is reduced. The metal atom always loses electrons and is oxidised.

4) If you ever need to predict whether or not a displacement reaction will happen, all you have to remember is that more reactive metals displace less reactive ones and you'll be fine and dandy.

5) You can also use this idea to write word or symbol equations for displacement reactions.

Rusting of Iron is a Redox Reaction

1) Iron and some steels will rust if they come into contact with air and water. Rusting only happens when the iron's in contact with both oxygen (from the air) and water.

2) Rust is a form of hydrated iron(III) oxide.

3) Here's the equation for rust:

> iron + oxygen + water → hydrated iron(III) oxide

4) Rusting of iron is a redox reaction.

5) This is why. Iron loses electrons when it reacts with oxygen. Each Fe atom loses three electrons to become Fe^{3+}. Iron's oxidised.

6) Oxygen gains electrons when it reacts with iron. Each O atom gains two electrons to become O^{2-}. Oxygen's reduced.

Remember OIL RIG.

Displacement reactions and rusting are types of redox reactions

Ahhhh, so that's what rust is... you learn something new everyday. Try writing some displacement reaction equations now — write the equation for the reaction between zinc and iron chloride ($FeCl_2$). What's being oxidised? What's being reduced? Practise till you can do it in your sleep.

Preventing Rust

Rust is a real pain — luckily there are some ways that you can stop it forming. Great news for bikes.

Metals are **Combined** with Other Things to **Prevent Rust**

1) Iron can be prevented from rusting by mixing it with <u>other metals</u> to make alloys.
2) <u>Steels</u> are alloys of iron with <u>carbon</u> and small quantities of other metals.
3) One of the most common steels is <u>stainless steel</u> — a rustproof alloy of iron, carbon and <u>chromium</u>.

Oil, Grease and Paint **Prevent Rusting**

You can <u>prevent rusting</u> by coating the iron with a <u>barrier</u>. This <u>keeps out the water</u>, <u>oxygen</u> or <u>both</u>.

1) <u>Painting</u> is ideal for large and small structures. It can also be nice and <u>colourful</u>.
2) <u>Oiling</u> or <u>greasing</u> has to be used when <u>moving parts</u> are involved, like on <u>bike chains</u>.

A Coat of **Tin** Can **Protect Steel from Rust**

1) <u>Tin plating</u> is where a coat of tin is applied to the object, e.g. food cans.
2) The tin acts as a <u>barrier</u>, stopping water and oxygen in the air from reaching the <u>surface</u> of the iron.
3) This only works as long as the <u>tin remains intact</u>. If the tin is <u>scratched</u> to reveal some iron, the <u>iron will lose electrons</u> in <u>preference</u> to the tin and the iron will rust even faster than if it was on its own.
4) That's why it's <u>not</u> always a good idea to buy the <u>reduced bashed tins</u> of food at the supermarket. They could be starting to <u>rust</u>.

More Reactive Metals Can Also be Used to **Prevent Iron Rusting**

You can also prevent rusting using the <u>sacrificial</u> method. You place a <u>more reactive metal</u> with the iron. The water and oxygen then react with this "sacrificial" metal instead of with the iron.

1) <u>Galvanising</u> is where a coat of <u>zinc</u> is put onto the object. The zinc acts as sacrificial protection — it's <u>more reactive</u> than iron so it'll <u>lose electrons in preference</u> to iron. The zinc also acts as a barrier. Steel <u>buckets</u> and <u>corrugated iron roofing</u> are often galvanised.

   Galvanising protects the metal underneath even when the zinc gets scratched.

2) Blocks of metal, e.g. <u>magnesium</u>, can be bolted to the iron. Magnesium will <u>lose electrons in preference to iron</u>. It's used on the hulls of <u>ships</u>, or on <u>underground iron pipes</u>.

<u>Don't get confused</u> about sacrificial protection — it's <u>not a displacement reaction</u>. There isn't a metal reacting with a metal salt — oxygen's reacting with a more reactive metal instead of a less reactive one.

Using coatings, plating and 'sacrificial metals' can prevent rusting

<u>Rust</u> is really annoying. It eats your bike, your car, your ship... Fortunately this page tells you how to prevent rusting, which means that you don't have to drive round in a car with big holes in it... Hooray.

Warm-Up and Exam Questions

There were a few tricky things on those last few pages, so here's your chance to check what you learnt.

Warm-Up Questions

1) Are nitrates soluble or insoluble?
2) Write down the word equation for the precipitation reaction between barium chloride and sodium sulfate.
3) Explain what you would do to make a dry sample of a soluble salt from an insoluble base.
4) What happens to electrons in a redox reaction?
5) What is the chemical name for rust?
6) What material are food cans often coated with to prevent rusting?
7) What metal is often used on the hulls of ships to prevent rusting?

Exam Questions

1   Jenny wanted to make a dry sample of silver chloride, AgCl, by precipitation.

   (a)   What property must a salt have to be made by precipitation?

   (1 mark)

   (b)   Jenny looked up the solubilities of some compounds she might use.

Compound	Formula	Solubility
silver oxide	Ag_2O	insoluble
silver nitrate	$AgNO_3$	soluble
silver carbonate	$AgCO_3$	insoluble
sulfuric acid	H_2SO_4	soluble
nitric acid	HNO_3	soluble
hydrochloric acid	HCl	soluble

   Write down a word equation using substances from the table that she could use to make silver chloride by precipitation.

   (1 mark)

   (c)   Outline the steps needed to give a pure dry sample of silver chloride after mixing the solutions.

   (3 marks)

2   Sacrificial methods such as galvanising can be used to prevent iron from rusting.

   (a)   Write down the word equation for the formation of rust.

   (1 mark)

   (b)   Explain why rusting is a redox reaction.

   (4 marks)

   (c)   Explain how galvanising prevents rusting.

   (2 marks)

   (d)   Explain how tin plating prevents iron from rusting.

   (1 mark)

Revision Summary for Section 8

Have a go at these questions and see how much you can remember. If you're not sure about any of them, don't just skulk past them, check back to the relevant page, get it in your head and tackle the question again. It's the best way, I promise. Especially when you're trying to learn a section full of equations and reactions like that one.

1) Give the meaning of this symbol: ☠

2) Write down the state symbol that means 'dissolved in water'.

3) What does the pH scale show?

4) What type of ions are always present in a) acids and b) alkalis?

5) What is neutralisation? Write down the general equation for neutralisation in terms of ions.

6) What is the difference between the strength of an acid and its concentration?

7) Name a metal that doesn't react at all with dilute acids.

8) What type of salts do hydrochloric acid and sulfuric acid produce?

9) What type of reaction is "acid + metal oxide", or "acid + metal hydroxide"?

10) Suggest a suitable acid and a suitable metal oxide/hydroxide to mix to form the following salts.
   a) copper chloride   b) calcium nitrate   c) zinc sulfate
   d) magnesium nitrate   e) sodium sulfate   f) potassium chloride

11) Iron chloride can made by mixing iron hydroxide (an insoluble base) with hydrochloric acid. Describe the method you would use to produce pure, solid iron chloride in the lab.

12) How can you tell when a neutralisation reaction is complete if both the base and the salt are soluble in water?

13) Fill in the gaps: A loss of electrons is _____. A gain of electrons is _____.

14) Give a symbol half-equation for the oxidation of Fe^{2+} to Fe^{3+}.

15) What is a displacement reaction?

16) The rusting of iron is a redox reaction.
    When rusting occurs, which substance is being reduced? Which substance is being oxidised?

17) Explain how greasing and painting protect against rust.

18) Why isn't it always a good plan to buy dented cans of beans?

19) An oil drilling platform uses sacrificial protection. What's "sacrificial protection"?

Rate of Reaction

Reactions can be <u>fast</u> or <u>slow</u> — you've probably already realised that. This page is about what affects the <u>rate of a reaction</u>, and the next page tells you what you can do to <u>measure it</u>.

Reactions Can Go At All Sorts of **Different Rates**

1) One of the <u>slowest</u> is the <u>rusting</u> of iron.
2) A <u>moderate speed</u> reaction is a <u>metal</u> (like magnesium) reacting with <u>acid</u> to produce a gentle stream of <u>bubbles</u>.
3) A <u>really fast</u> reaction is an <u>explosion</u>, where it's all over in a <u>fraction</u> of a second.

The **Rate of a Reaction** Depends on **Four Things**:

1) Temperature

2) Concentration — (or <u>pressure</u> for gases)

3) Catalyst

4) Surface area of solids — (or <u>size</u> of solid pieces)

Typical Graphs for Rate of Reaction

The plot below shows how the rate of a particular reaction varies under <u>different conditions</u>. The <u>quickest reaction</u> is shown by the line with the <u>steepest slope</u>. Also, the faster a reaction goes, the sooner it finishes, which means that the line becomes <u>flat</u> earlier.

1) <u>Graph 1</u> represents the original <u>fairly slow</u> reaction. The graph is not too steep.

2) <u>Graphs 2 and 3</u> represent the reaction taking place <u>quicker</u> but with the <u>same initial amounts</u>. The slope of the graphs gets steeper.

3) The <u>increased rate</u> could be due to <u>any</u> of these:

   a) increase in <u>temperature</u>
   b) increase in <u>concentration</u> (or pressure)
   c) <u>catalyst</u> added
   d) solid reactant crushed up into <u>smaller bits</u>.

You could also show the amount of reactant used up over time instead — the graphs would have the same shape.

Amount of product evolved

④ faster, and more reactants

end of reaction

③ much faster reaction

② faster reaction

① original reaction

Time

4) <u>Graph 4</u> produces <u>more product</u> as well as going <u>faster</u>. This can <u>only</u> happen if <u>more reactant(s)</u> are added at the start. <u>Graphs 1, 2 and 3</u> all converge at the same level, showing that they all produce the same amount of product, although they take <u>different</u> times to get there.

A steep graph means a speedy reaction

<u>Industrial</u> reactions generally use a <u>catalyst</u> and are done at <u>high temperature and pressure</u>. Time is money, so the faster an industrial reaction goes the better... but only <u>up to a point</u>. Chemical plants are quite expensive to rebuild if they get blown into lots and lots of teeny tiny pieces.

Measuring Rates of Reaction

If you want to know the rate of reaction then it's fairly easy to <u>measure</u> it.
Here are <u>three</u> ways of measuring the rate of a reaction for you to look at...

Ways to **Measure the Rate** of a Reaction

The <u>rate of a reaction</u> can be observed <u>either</u> by measuring how quickly the reactants are used up
or how quickly the products are formed. It's usually a lot easier to measure <u>products forming</u>.

The rate of reaction can be calculated using the following formula:

$$\text{Rate of reaction} = \frac{\text{amount of reactant used or amount of product formed}}{\text{time}}$$

There are different ways that the rate of a reaction can be <u>measured</u>. Here are three:

1) Precipitation

1) This is when the product of the reaction is a <u>precipitate</u> which <u>clouds</u> the solution.

2) Observe a <u>mark</u> through the solution and measure how long it takes for it to <u>disappear</u>.

3) The <u>quicker</u> the mark disappears, the <u>quicker</u> the reaction.

4) This only works for reactions where the initial solution is rather <u>see-through</u>.

5) The result is very <u>subjective</u> — <u>different people</u> might not agree
    over the <u>exact</u> point when the mark 'disappears'.

Measuring Rates of Reaction

2) *Change in Mass (Usually Gas Given Off)*

1) Measuring the speed of a reaction that <u>produces a gas</u> can be carried out on a <u>mass balance</u>.

2) As the gas is released the mass <u>disappearing</u> is easily measured on the balance.

3) The <u>quicker</u> the reading on the balance <u>drops</u>, the <u>faster</u> the reaction.

4) <u>Rate of reaction graphs</u> are particularly easy to plot using the results from this method.

5) This is the <u>most accurate</u> of the three methods described because the mass balance is very accurate. But it has the <u>disadvantage</u> of releasing the gas straight into the room.

3) *The* **Volume** *of Gas Given Off*

1) This involves the use of a <u>gas syringe</u> to measure the <u>volume</u> of gas given off.

2) The <u>more</u> gas given off during a given <u>time interval</u>, the <u>faster</u> the reaction.

3) A graph of <u>gas volume</u> against <u>time elapsed</u> could be plotted to give a rate of reaction graph.

4) Gas syringes usually give volumes accurate to the <u>nearest millilitre</u>, so they're quite accurate. You have to be quite careful though — if the reaction is too <u>vigorous</u>, you can easily blow the plunger out of the end of the syringe.

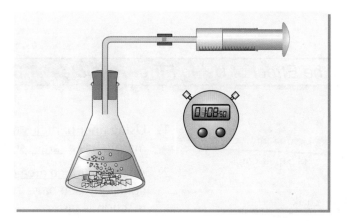

Each of these methods has pros and cons

The mass balance method is only accurate as long as the flask isn't too hot, otherwise you lose mass by <u>evaporation</u> as well as in the reaction. The first method <u>isn't</u> very accurate, but if you're not producing a gas you can't use either of the other two. Ah well.

Rate of Reaction Experiments

Remember: Any reaction can be used to investigate any of the four factors that affect the rate.
The next four pages illustrate four important reactions, but only one factor is considered for each.
But you can just as easily use, say, the marble chips/acid reaction to test the effect of temperature instead.

1) Reaction of Hydrochloric Acid and Marble Chips

This experiment is often used to demonstrate the effect of breaking the solid up into small bits.

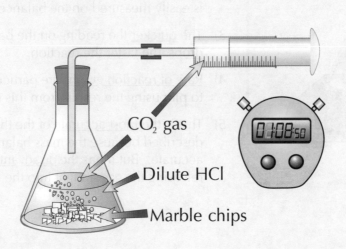

CO₂ gas

Dilute HCl

Marble chips

1) Measure the volume of gas evolved with a gas syringe and take readings at
    regular intervals.

2) Make a table of readings and plot them as a graph. You choose regular
    time intervals, and time goes on the x-axis and volume goes on the y-axis.

3) Repeat the experiment with exactly the same volume of acid, and exactly
    the same mass of marble chips, but with the marble more crunched up.

4) Then repeat with the same mass of powdered chalk instead of marble chips.

This Graph Shows the Effect of Using Finer Particles of Solid

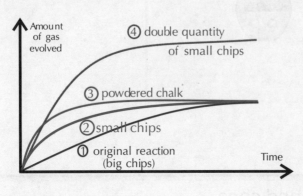

1) Using finer particles means that the
    marble has a larger surface area.

2) A larger surface area causes more frequent
    collisions (see page 168) so the rate of
    reaction is faster.

3) Line 4 shows the reaction if a greater mass
    of small marble chips is added. The extra
    surface area gives a quicker reaction and
    there is also more gas evolved overall.

Rate of Reaction Experiments

The reaction of <u>magnesium metal</u> with <u>dilute HCl</u> is often used to determine the effect of <u>concentration</u>.

2) Reaction of *Magnesium Metal With Dilute HCl*

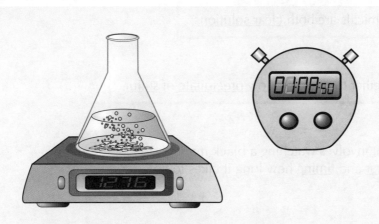

1) <u>This reaction</u> is good for measuring the effects of <u>increased concentration</u> (as is the marble/acid reaction).

2) This reaction gives off <u>hydrogen gas</u>, which we can measure with a <u>mass balance</u>, as shown.

3) In this experiment, <u>time</u> also goes on the <u>x-axis</u> and <u>volume</u> goes on the <u>y-axis</u>.
   (The other method is to use a gas syringe, see page 162.)

This Graph Shows the Effect of Using *More Concentrated Acid Solutions*

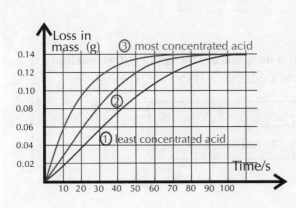

1) Take <u>readings</u> of mass at <u>regular</u> time intervals.

2) Put the results in a <u>table</u> and work out the <u>loss in mass</u> for each reading. <u>Plot a graph</u>.

3) <u>Repeat</u> with <u>more concentrated</u> acid solutions, but always with the <u>same</u> amount of magnesium.

4) The <u>volume</u> of acid must always be kept <u>the same</u> too — only the <u>concentration</u> is increased.

5) The three graphs show the <u>same</u> old pattern — a <u>higher</u> concentration giving a <u>steeper graph</u>, with the reaction <u>finishing</u> much quicker.

Rate of Reaction Experiments

The effect of <u>temperature</u> on the rate of a reaction can be measured using a <u>precipitation</u> reaction.

3) *Sodium Thiosulfate* and *HCl* Produce a *Cloudy Precipitate*

1) These two chemicals are both <u>clear solutions</u>.

2) They react together to form a <u>yellow precipitate</u> of <u>sulfur</u>.

3) The experiment involves watching a black mark <u>disappear</u> through the <u>cloudy sulfur</u> and <u>timing</u> how long it takes to go.

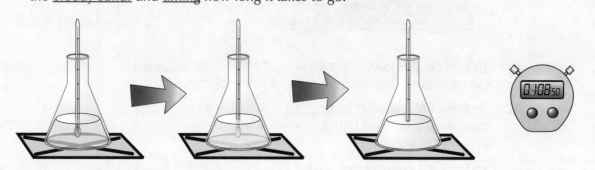

4) The reaction can be <u>repeated</u> for solutions at different <u>temperatures</u>. In practice, that's quite hard to do accurately and safely (it's not a good idea to heat an acid directly). The best way to do it is to use a <u>water bath</u> to heat both solutions to the right temperature <u>before you mix them</u>.

5) The <u>depth</u> of liquid must be kept the <u>same</u> each time, of course.

6) The results will of course show that the <u>higher</u> the temperature the <u>quicker</u> the reaction and therefore the <u>less time</u> it takes for the mark to <u>disappear</u>. These are typical results:

Temperature (°C)	20	25	30	35	40
Time taken for mark to disappear (s)	193	151	112	87	52

This reaction can <u>**also**</u> be used to test the effects of <u>concentration</u>.

This reaction <u>**doesn't**</u> give a set of graphs. All you get is a set of <u>readings</u> of how long it took till the mark disappeared for each temperature.

Rate of Reaction Experiments

Good news — this is the last rate experiment. This one looks at how a <u>catalyst</u> affects rate of reaction.

4) The **Decomposition** of **Hydrogen Peroxide**

This is a <u>good</u> reaction for showing the effect of different <u>catalysts</u>. The decomposition of hydrogen peroxide is:

$$2H_2O_{2 \, (aq)} \rightleftharpoons 2H_2O_{(l)} + O_{2 \, (g)}$$

1) This is normally quite <u>slow</u> but a sprinkle of <u>manganese(IV) oxide catalyst</u> speeds it up no end. Other catalysts which work are found in:
a) <u>potato peel</u> and b) <u>blood</u>.

2) <u>Oxygen gas</u> is given off, which provides an <u>ideal way</u> to measure the rate of reaction using the <u>gas syringe</u> method.

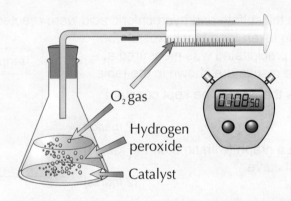

O$_2$ gas

Hydrogen peroxide

Catalyst

This Graph Shows the Effect of Using **Different Catalysts**

1) Same old graphs of course.

2) <u>Better</u> catalysts give a <u>quicker reaction</u>, which is shown by a <u>steeper graph</u> which levels off quickly.

3) This reaction can also be used to measure the effects of <u>temperature</u>, or of <u>concentration</u> of the H_2O_2 solution. The graphs will look just the same.

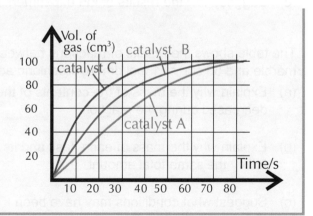

Blood is a catalyst? — eeurgh...

This stuff's about comparing those pretty rate of reaction graphs. They tell you the amount of <u>product</u> made (or reactant used up) and the <u>rate of reaction</u>. The <u>steeper</u> the curve, the <u>faster</u> the reaction.

Warm-Up and Exam Questions

Time to test your knowledge again. This time on the rates of chemical reactions. If you struggle with these questions and you don't feel up to speed, it's time to have another look at the last few pages.

Warm-Up Questions

1) Give an example of a reaction that happens very slowly, and one that is very fast.
2) Give three ways of increasing the rate of a reaction between magnesium and sulfuric acid.
3) How would reducing the concentration of an acid affect the time taken for a piece of zinc to react with it?
4) Oxidation of lactose in milk makes it go 'sour'. How could this reaction be slowed down?
5) Describe one way of monitoring a reaction in which a gas is given off.

Exam Questions

1 Set volumes of sodium thiosulfate and hydrochloric acid were reacted at different temperatures. The time taken for a black cross to be obscured by the sulfur precipitated was measured at each temperature. The results are shown in the table.

Temperature (°C)	Time (s)
55	6
36	11
24	17
16	27
9	40
5	51

(a) Give two variables that should be kept constant in this experiment.

(2 marks)

(b) Plot the results on a graph (with time on the x-axis) and draw a best-fit curve.

(2 marks)

(c) Describe the relationship illustrated by your graph.

(1 mark)

(d) Describe how the results would change if the sodium thiosulfate concentration was reduced.

(2 marks)

(e) Suggest how the results of the experiment could be made more reliable.

(1 mark)

2 The table shows the results of reactions between excess marble and 50 cm³ of 1 mol/dm³ hydrochloric acid.

Time (min)	Mass of flask A (g)	Mass of flask B (g)
0	121.6	121.6
1	120.3	119.8
2	119.7	119.2
3	119.4	119.1
4	119.2	119
5	119.1	119
6	119	119
7	119	

(a) Explain why the mass of the contents of the flasks decreased during the reaction.

(1 mark)

(b) Explain why the mass of each flask and its contents fell by the same total amount.

(1 mark)

(c) Suggest what conditions may have been different inside flask B.

(1 mark)

(d) In both reactions, the rate is fastest at the beginning. Suggest why.

(1 mark)

Collision Theory

Reaction rates are explained by collision theory. It's really simple.

1) Collision theory just says that the rate of a reaction simply depends on how often and how hard the reacting particles collide with each other.

2) The basic idea is that particles have to collide in order to react, and they have to collide hard enough (with enough energy).

More Collisions Increases the Rate of Reaction

The effects of temperature, concentration and surface area on the rate of reaction can be explained in terms of how often the reacting particles collide successfully.

1) HIGHER TEMPERATURE Increases Collisions

When the temperature is increased the particles all move quicker. If they're moving quicker, they're going to collide more often.

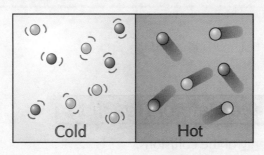

Cold   Hot

2) HIGHER CONCENTRATION (or PRESSURE) Increases Collisions

If a solution is made more concentrated it means there are more particles of reactant knocking about between the water molecules which makes collisions between the important particles more likely.

In a gas, increasing the pressure means the particles are more squashed up together so there will be more frequent collisions.

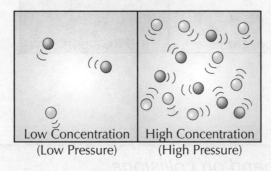

Low Concentration   High Concentration
(Low Pressure)   (High Pressure)

Collision Theory

3) LARGER SURFACE AREA Increases Collisions

If one of the reactants is a <u>solid</u> then <u>breaking it up</u> into <u>smaller</u> pieces will <u>increase the total surface area</u>. This means the particles around it in the solution will have <u>more area to work on</u>, so there'll be <u>more frequent collisions</u>.

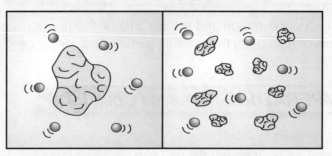

Small surface area   Big surface area

Faster collisions Increase the Rate of Reaction

<u>Higher temperature</u> also increases the <u>energy</u> of the collisions, because it makes all the particles <u>move faster</u>.

Increasing the <u>temperature</u> causes <u>faster collisions</u>.

Reactions <u>only happen</u> if the particles collide with <u>enough energy</u>.

The <u>minimum amount</u> of energy needed by the particles to react is known as the <u>activation energy</u>.

At a <u>higher temperature</u> there will be <u>more particles</u> colliding with <u>enough energy</u> to make the reaction happen.

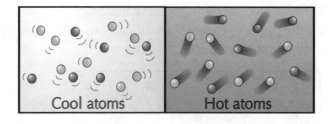

Cool atoms   Hot atoms

Chemical reactions depend on collisions

Once you've read everything on this page, the rates of reaction stuff should start making <u>a lot more sense</u> to you. The concept's fairly simple — the <u>more often</u> particles bump into each other, and the <u>harder</u> they hit when they do, the <u>faster</u> the reaction happens.

Catalysts

In industrial reactions, the main thing they're interested in is making a <u>nice profit</u>.
Catalysts are helpful for this — they can reduce costs and increase the amount of product.

A **Catalyst** Increases the **Number of Successful Collisions**

1) A <u>catalyst</u> is a substance which increases the <u>speed of a reaction</u>, <u>without</u> being chemically changed or used up in the reaction — and because it isn't used up, you only need a <u>tiny bit</u> of it to catalyse large amounts of reactants.

2) Catalysts tend to be very <u>fussy</u> about which reactions they catalyse though — you can't just stick any old catalyst in a reaction and expect it to work.

3) A catalyst works by giving the reacting particles a <u>surface</u> to stick to where they can bump into each other — and <u>reduces the energy needed</u> by the particles before they react.

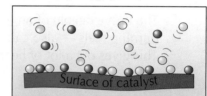
Surface of catalyst

4) So the <u>overall number</u> of collisions isn't increased, but the number of <u>successful collisions</u> is.

Catalysts Help **Reduce Costs** in Industrial Reactions

1) Catalysts are <u>very important</u> for <u>commercial reasons</u> — most industrial reactions use them.

2) <u>Catalysts</u> increase the rate of the reaction, which saves a lot of <u>money</u> simply because the plant doesn't need to operate for <u>as long</u> to produce the <u>same amount</u> of stuff.

3) Alternatively, a catalyst will allow the reaction to work at a <u>much lower temperature</u>. That reduces the <u>energy</u> used up in the reaction (the <u>energy cost</u>), which is good for <u>sustainable development</u> (see page 141) and can save a lot of money too.

4) There are <u>disadvantages</u> to using catalysts, though.

5) They can be very expensive to buy, and often need to be removed from the product and cleaned. They never get <u>used up</u> in the reaction though, so once you've got them you can use them <u>over and over</u> again.

6) Different <u>reactions</u> use different <u>catalysts</u>, so if you make <u>more than one product</u> at your plant, you'll probably need to buy different catalysts for them.

7) Catalysts can be 'poisoned' by impurities, so they <u>stop working</u>, e.g. sulfur impurities can poison the iron catalyst used in the Haber process (used to make ammonia for fertilisers). That means you have to keep your reaction mixture very <u>clean</u>.

A big advantage of catalysts is that they can be used over and over

And they're not only used in <u>industry</u>... every useful chemical reaction in the human body is catalysed by a <u>biological catalyst</u> (an enzyme). If the reactions in the body were just left to their own devices, they'd take so long to happen, we couldn't exist. Quite handy then, these catalysts.

Energy Transfer in Reactions

Whenever chemical reactions occur <u>energy</u> is <u>transferred to</u> or <u>from</u> the <u>surroundings</u>.

In an **Exothermic** Reaction, Heat is **Given Out**

An <u>EXOTHERMIC reaction</u> is one which <u>transfers energy</u> to the surroundings, usually in the form of <u>heat</u> and usually shown by a <u>rise in temperature</u>.

1) **Burning** Fuels

The best example of an <u>exothermic</u> reaction is <u>burning fuels</u> — also called <u>COMBUSTION</u>. This gives out a lot of heat — it's very exothermic.

2) **Neutralisation** Reactions

<u>Neutralisation reactions</u> (acid + alkali) are also exothermic — see page 145.

ACID

<u>Don't</u> do it like this!

ALKALI

3) **Oxidation** Reactions

Many <u>oxidation reactions</u> are exothermic. For example:

Adding sodium to water <u>produces heat</u>, so it must be <u>exothermic</u> — see page 110. The sodium emits <u>heat</u> and moves about on the surface of the water as it is oxidised.

Exothermic reactions have lots of <u>everyday uses</u>. For example, some <u>hand warmers</u> use the exothermic <u>oxidation of iron</u> in air (with a salt solution catalyst) to generate <u>heat</u>. <u>Self heating cans</u> of hot chocolate and coffee also rely on exothermic reactions between <u>chemicals</u> in their bases.

Energy Transfer in Reactions

In an *Endothermic* Reaction, Heat is *Taken In*

An <u>ENDOTHERMIC reaction</u> is one which <u>takes in energy</u> from the surroundings, usually in the form of <u>heat</u> and is usually shown by a <u>fall in temperature</u>.

Endothermic reactions are much <u>less common</u>. <u>Thermal decompositions</u> are a good example:

Heat must be supplied to make calcium carbonate <u>decompose</u> to make calcium oxide.

$$CaCO_3 \rightarrow CaO + CO_2$$

Endothermic reactions also have everyday uses. For example, some <u>sports injury packs</u> use endothermic reactions — they <u>take in heat</u> and the pack becomes very <u>cold</u>. More <u>convenient</u> than carrying ice around.

Reversible Reactions Can Be *Endothermic* and *Exothermic*

In reversible reactions (see page 141), if the reaction is <u>endothermic</u> in <u>one direction</u>, it will be <u>exothermic</u> in the <u>other direction</u>. The <u>energy absorbed</u> by the endothermic reaction is <u>equal</u> to the <u>energy released</u> during the exothermic reaction.

A good example is the <u>thermal decomposition of hydrated copper sulfate</u>.

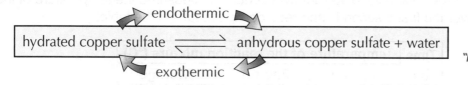

endothermic

hydrated copper sulfate ⇌ anhydrous copper sulfate + water

exothermic

"Anhydrous" just means "without water", and "hydrated" means "with water".

1) If you <u>heat blue hydrated</u> copper(II) sulfate crystals it drives the water off and leaves <u>white anhydrous</u> copper(II) sulfate powder. This is endothermic.

Water vapour

2) If you then <u>add</u> a couple of drops of <u>water</u> to the <u>white powder</u> you get the <u>blue crystals</u> back again. This is exothermic.

Right, so burning gives out heat — really...

This whole energy transfer thing is a fairly simple idea — don't be put off by the long words. Remember, "<u>exo-</u>" = <u>exit</u>, "<u>-thermic</u>" = <u>heat</u>, so an exothermic reaction is one that <u>gives out</u> heat. And "<u>endo-</u>" = erm... the other one. Okay, so there's no easy way to remember that one. Tough.

Warm-Up and Exam Questions

Here are some more questions to have a go at. If you can't do these ones then you won't be able to do the ones in the exam either. And you don't want that. If you're struggling, read the pages over again.

Warm-Up Questions

1) According to collision theory, what must happen in order for two particles to react?
2) Why does an increase in concentration of solutions increase the rate of a reaction?
3) Give a definition of a catalyst.
4) An endothermic reaction happens when ammonium nitrate is dissolved in water. Predict how the temperature of the solution will change during the reaction.

Exam Questions

1   *In this question you will be assessed on the quality of your English, the organisation of your ideas and your use of appropriate specialist vocabulary.*

Hydrogen and ethene react to form ethane. Nickel can be used as a catalyst for this reaction.

Using your knowledge of collision theory, suggest how the rate of this reaction can be increased.

(6 marks)

2   A student added hydrochloric acid to sodium hydroxide. He measured the temperature of the reaction mixture over the first 5 seconds and recorded his results in the table.

Time (s)	Temperature of the reaction mixture (°C)		
	1st run	**2nd run**	**Average**
0	22.0	22.0	
1	25.6	24.4	
2	28.3	28.1	
3	29.0	28.6	
4	28.8	28.8	
5	28.3	28.7	

(a)   State the name given to this type of reaction.

(1 mark)

(b)   Complete the table by calculating the average temperature of the reaction mixture during the two runs.

(2 marks)

(c)   Calculate the maximum average increase in temperature during the reaction.

(1 mark)

(d)   Is this reaction exothermic or endothermic? Explain your answer.

(2 marks)

Energy

Whenever chemical reactions occur, there are changes in <u>energy</u>. Changes in energy during a chemical reaction can be explained by <u>making bonds</u> or <u>breaking bonds</u>.

*Energy Must Always be **Supplied** to **Break Bonds**...*
*...and Energy is Always **Released** When **Bonds Form***

1) During a chemical reaction, <u>old bonds</u> are <u>broken</u> and <u>new bonds</u> are <u>formed</u>.

2) Energy must be <u>supplied</u> to break <u>existing bonds</u> — so bond breaking is an <u>endothermic</u> process. Energy is <u>released</u> when new bonds are <u>formed</u> — so bond formation is an <u>exothermic</u> process.

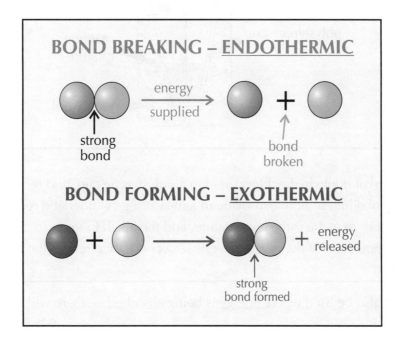

3) In an <u>endothermic</u> reaction, the energy <u>required</u> to break old bonds is <u>greater</u> than the energy <u>released</u> when <u>new bonds</u> are formed.

4) In an <u>exothermic</u> reaction, the energy <u>released</u> in bond formation is <u>greater</u> than the energy used in <u>breaking</u> old bonds.

Save energy — break fewer bonds...

You can get <u>cooling packs</u> that use an <u>endothermic</u> reaction to draw heat from an injury. The pack contains two compartments with different chemicals in. When you use it, you snap the partition and the chemicals <u>mix</u> and <u>react</u>, taking in <u>heat</u> — pretty cool, I reckon (no pun intended).

Energy

The <u>fuels</u> we use are great because they release <u>loads of energy</u> when they burn.
But they cause a few <u>problems</u> as well — none bigger than <u>global warming</u>.

Energy transfer can be measured

1) You can measure the amount of <u>energy released</u> by a <u>chemical reaction</u> (in solution) by taking the <u>temperature of the reagents</u> (making sure they're the same), <u>mixing</u> them in a <u>polystyrene cup</u> and measuring the <u>temperature of the solution</u> at the <u>end</u> of the reaction. Easy.

2) The biggest <u>problem</u> with energy measurements is the amount of energy <u>lost to the surroundings</u>.

3) You can reduce it a bit by putting the polystyrene cup into a <u>beaker of cotton wool</u> to give <u>more insulation</u>, and putting a <u>lid</u> on the cup to reduce energy lost by <u>evaporation</u>.

4) This method works for reactions of <u>solids with water</u> (e.g. dissolving ammonium nitrate in water) as well as for <u>neutralisation</u> reactions.

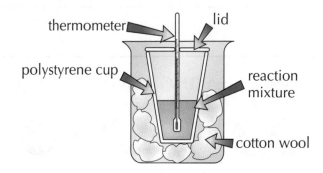

Diagram labels: thermometer, lid, polystyrene cup, reaction mixture, cotton wool

<u>Example:</u>
1) Place 25 cm³ of dilute hydrochloric acid in a polystyrene cup, and record its temperature.
2) Put 25 cm³ of dilute sodium hydroxide in a measuring cylinder and record its temperature.
3) As long as they're at the same temperature, add the alkali to the acid and stir.
4) Take the temperature of the mixture every 30 seconds, and record the highest temperature it reaches.

5) This method can also be used where <u>energy</u> is being <u>absorbed</u> — there will be a <u>fall</u> in temperature.

*Fuels provide **energy** — but there are **consequences***

1) Fuels release <u>energy</u> which we use in loads of ways — e.g. to generate electricity and to power cars.
2) Burning fuels has various effects on the <u>environment</u>. For example, burning fossil fuels releases CO_2, a greenhouse gas. This causes <u>global warming</u> and other types of <u>climate change</u>.
3) It'll be <u>expensive</u> to slow down these effects, and to put things right. Developing alternative energy sources (e.g. tidal power) costs money.
4) Crude oil is <u>running out</u>. We use <u>a lot of fuels</u> made from crude oil (e.g. <u>petrol and diesel</u>) and as it runs out it will get more expensive. This means that everything that's <u>transported</u> by lorry, train or plane gets more expensive too. So the <u>price of crude oil</u> has a big economic effect.

Crude oil — using it is bad, but we can't do without it

There's no hiding from it — finding the fuels to produce enough power for everyone is a <u>huge challenge</u>. Especially trying to do it without trashing the <u>environment</u> in the process. It's a problem alright. And it's likely to become even more of a problem in the future when crude oil starts to <u>run out</u>...

Energy and Fuels

Burning <u>fuels</u> releases <u>energy</u>. Just how much energy you can find using <u>calorimetry</u>. Bet you can't wait.

*Fuel energy is **calculated** using **calorimetry***

Different fuels produce <u>different amounts of energy</u>. To measure the amount of energy released when a fuel is burnt, you can simply burn the fuel and use the flame to <u>heat up some water</u>. Of course, this has to have a fancy chemistry name — <u>calorimetry</u>. Calorimetry uses a <u>glass</u> or <u>metal container</u> (it's usually made of <u>copper</u> because copper conducts heat so well).

Method:

1) Put 50 g of water in the copper can and <u>record its temperature</u>.

2) <u>Weigh the spirit burner</u> and lid.

3) Put the spirit burner underneath the can, and light the wick. Heat the water, <u>stirring constantly</u>, until the temperature reaches about <u>50 °C</u>.

4) <u>Put out the flame</u> using the burner lid, and measure the <u>final temperature</u> of the water.

5) <u>Weigh</u> the spirit burner and lid <u>again</u>.

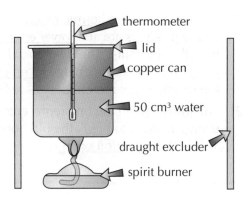

You can use pretty much the same method to calculate the amount of energy produced by <u>foods</u>. The only problem is that when you set food on fire, it tends to <u>go out</u> after a bit.

Example: to work out the energy per gram of methylated spirit (meths):

1) Mass of spirit burner + lid before heating = 68.75 g

2) Mass of spirit burner + lid after heating = 67.85 g

Mass of meths burnt = 0.9 g

3) Temperature of water in copper can before heating = 21.5 °C

4) Temperature of water in copper can after heating = 52.5 °C

Temperature change in 50 g of water due to heating = 31.0 °C

5) So 0.9 g of fuel produces enough energy to heat up 50 g of water by 31 °C.

6) It takes 4.2 joules of energy to heat up 1 g of water by 1 °C. This is known as the specific heat capacity of water. *You'll be told this in the exam if you need it.*

$$Q = mc\Delta T$$

ENERGY TRANSFERRED (in J)	=	MASS OF WATER (in g)	×	SPECIFIC HEAT CAPACITY OF WATER (= 4.2)	×	TEMPERATURE CHANGE (in °C)
Q		m		c		ΔT

7) Therefore, the energy produced in this experiment = 50 × 4.2 × 31 = <u>6510 joules</u>.

8) So 0.9 g of meths produces 6510 joules of energy... ... meaning 1 g of meths produces 6510/0.9 = <u>7233 J or 7.233 kJ</u>

Energy's wasted heating the can, air, etc. — so this figure will often be much lower than the <u>actual</u> energy content.

Energy from fuels — it's a burning issue...

So there you have it — how to find the amount of energy a fuel produces as it burns, from experiment to calculation. You might sometimes see energy values given in <u>calories</u> (1 calorie = 4.2 joules).

Bond Energies

Remember — chemical reactions involve a change in energy. Energy level diagrams show this change.

Energy Level Diagrams Show if it's *Exo-* or *Endo-thermic*

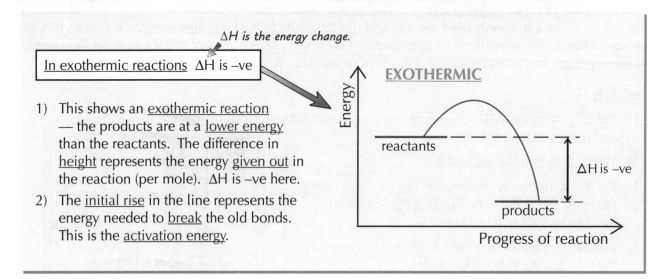

ΔH is the energy change.

In exothermic reactions ΔH is –ve

1) This shows an exothermic reaction — the products are at a lower energy than the reactants. The difference in height represents the energy given out in the reaction (per mole). ΔH is –ve here.

2) The initial rise in the line represents the energy needed to break the old bonds. This is the activation energy.

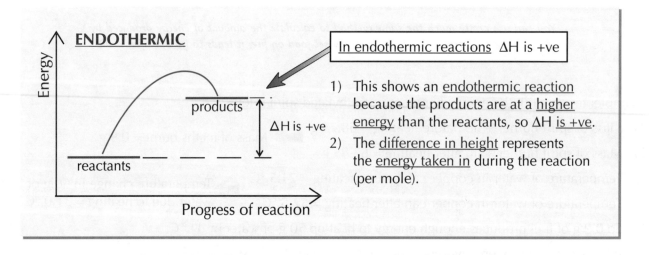

In endothermic reactions ΔH is +ve

1) This shows an endothermic reaction because the products are at a higher energy than the reactants, so ΔH is +ve.

2) The difference in height represents the energy taken in during the reaction (per mole).

The *Activation Energy* is *Lowered* by *Catalysts*

1) The activation energy represents the minimum energy needed by reacting particles to break their bonds.

2) A catalyst provides a different pathway for a reaction that has a lower activation energy (so the reaction happens more easily and more quickly).

3) This is represented by the lower curve on the diagram showing a lower activation energy.

4) The overall energy change for the reaction, ΔH, remains the same though.

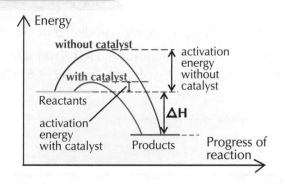

Bond Energies

You need to be able to <u>work out</u> ΔH for a particular reaction.

You Can Calculate *Energy Change* Using *Bond Energies*

1) <u>Every</u> chemical bond has a particular <u>bond energy</u> associated with it. This <u>bond energy</u> varies slightly depending on what <u>compound</u> the bond occurs in.

2) You can use these <u>known bond energies</u> to calculate the <u>overall energy change</u> for a reaction.

Example: *The Formation of HCl*

Using known bond energies you can <u>calculate</u> the <u>energy change</u> for this reaction:

$$H - H + Cl - Cl \rightarrow 2H - Cl$$
$$H_2 + Cl_2 \rightarrow 2HCl$$

The bond energies you need are:
- H—H: +436 kJ/mol
- Cl—Cl: +242 kJ/mol
- H—Cl: +431 kJ/mol

1) <u>Breaking</u> one mole of H—H and one mole of Cl—Cl bonds <u>requires</u>:
436 + 242 = <u>678 kJ</u>

2) <u>Forming</u> <u>two</u> moles of H—Cl bonds <u>releases</u> 2 × 431 = <u>862 kJ</u>

3) <u>Overall</u> more energy is <u>released</u> than is used to form the products:
862 – 678 = <u>184 kJ/mol</u> released.

4) Since this is energy <u>released</u>, if we wanted to show ΔH we'd need to put a <u>negative sign</u> in front of it to indicate that it's an <u>exothermic</u> reaction, like this:

$$\Delta H = -184 \text{ kJ/mol}$$

Energy change is down to what bonds are broken and made

I admit — it's a bit like maths, this. But think how many times you've heard <u>energy efficiency</u> mentioned over the last few years. Well, this kind of calculation is used in working out whether we're using resources efficiently or not. So even if it's not exciting, it's useful at least.

Equilibrium

Equilibrium is all to do with the rate of a reversible reaction...

> A reversible reaction is one where the products of the reaction can themselves react to produce the original reactants
>
> A + B $\rightleftharpoons$ C + D

The '$\rightleftharpoons$' shows the reaction goes both ways.

Reversible Reactions Will Reach Equilibrium

1) As the reactants (A and B) react, their concentrations fall — so the forward reaction will slow down. But as more and more products (C and D) are made and their concentrations rise, the backward reaction will speed up.

2) After a while the forward reaction will be going at exactly the same rate as the backward one — this is equilibrium.

3) At equilibrium both reactions are still happening, but there's no overall effect (it's a dynamic equilibrium). This means the concentrations of reactants and products have reached a balance and won't change.

4) Equilibrium is only reached if the reversible reaction takes place in a 'closed system'. A closed system just means that none of the reactants or products can escape.

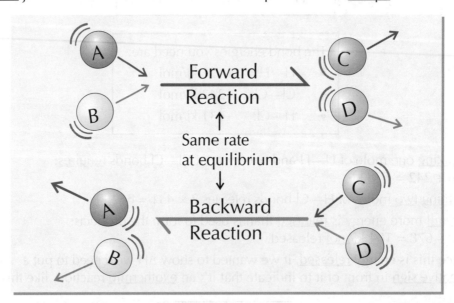

The Position of Equilibrium Can be on the Right or the Left

When a reaction's at equilibrium it doesn't mean the amounts of reactants and products are equal.

1) Sometimes the equilibrium will lie to the right — this basically means "lots of the products and not much of the reactants" (i.e. the concentration of product is greater than the concentration of reactant).

2) Sometimes the equilibrium will lie to the left — this basically means "lots of the reactants but not much of the products" (the concentration of reactant is greater than the concentration of product).

3) The exact position of equilibrium depends on the conditions (as well as the reaction itself).

Changing Equilibrium

Equilibrium is all to do with the <u>rate</u> of a <u>reversible</u> reaction...

Three Things Can Change the Position of Equilibrium:

1) <u>TEMPERATURE</u>
2) <u>PRESSURE</u> (only affects equilibria involving gases)
3) <u>CONCENTRATION</u>

1 equilibrium, but 2 equilibria.

See below for <u>why</u> these things affect the equilibrium position.
But now's a good time to make a mental note of this potential elephant trap...

Adding a CATALYST doesn't change the equilibrium position:

1) Catalysts speed up <u>both</u> the <u>forward</u> and <u>backward</u> reactions by the <u>same amount</u>.

2) So, adding a catalyst means the reaction reaches equilibrium <u>quicker</u>, but you end up with the <u>same amount</u> of product as you would without the catalyst.

The Equilibrium Tries to Minimise Any Changes You Make

Temperature

All reactions are <u>exothermic</u> in one direction and <u>endothermic</u> in the other (see p. 171).
1) If you <u>decrease</u> the temperature, the equilibrium will move to try and <u>increase</u> it — the equilibrium moves in the <u>exothermic direction</u> to produce more heat.
2) If you <u>raise</u> the temperature, the equilibrium will move to try and <u>decrease</u> it — the equilibrium moves in the <u>endothermic direction</u>.

$$N_2 + 3H_2 \rightleftharpoons 2NH_3$$

The forward reaction is exothermic — a decrease in temperature moves the equilibrium to the right (more products).

Pressure

Changing this only affects an equilibrium involving <u>gases</u>.
1) If you <u>increase</u> the pressure, the equilibrium tries to <u>reduce</u> it — the equilibrium moves in the <u>direction</u> where there are fewer moles of gas.
2) If you <u>decrease</u> the pressure, the equilibrium tries to <u>increase</u> it — it moves in the <u>direction</u> where there are more moles of gas.

$$N_2 + 3H_2 \rightleftharpoons 2NH_3$$

There are 4 moles on the left, but only 2 on the right. So, if you increase the pressure, the equilibrium shifts to the right.

Concentration

Same reaction again... $N_2 + 3H_2 \rightleftharpoons 2NH_3$

1) If you <u>increase the concentration</u> of the reactants by adding more N_2 or H_2, the equilibrium tries to decrease it by shifting to the <u>right</u> (making <u>more NH_3</u>).

2) If you <u>increase the concentration</u> of product by adding more NH_3, the equilibrium tries to reduce it again by shifting to the <u>left</u> (making <u>more N_2 and H_2</u>).

If you decrease the concentration of N_2, H_2 or NH_3 by removing them, the equilibrium moves to try and increase the concentration again.

Changing Equilibrium

It's important that you can <u>interpret</u> any <u>data</u> you might get in an exam about equilibria.

Make Sure You Can Read Equilibrium Tables and Graphs

You might be asked to <u>interpret data</u> about <u>equilibrium</u>, so you'd better know what you're doing. The Haber process (see page 229) is a great example of all this...

$$N_2 + 3H_2 \rightleftharpoons 2NH_3$$

The forward reaction is exothermic.

First off, a table...

Pressure (atmospheres)	100	200	300	400	500
% of ammonia at 450 °C	14	26	34	39	42

As the <u>pressure increases</u>, the proportion of ammonia <u>increases</u> (exactly what you'd expect — since increasing the pressure shifts the equilibrium to the side with fewer moles of gas — here, the right).

And now a graph...

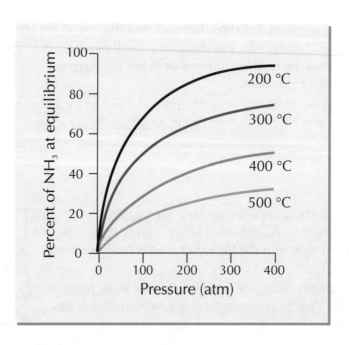

1) This time, each different line represents a different <u>temperature</u>.

2) As the temperature <u>increases</u>, the proportion of ammonia <u>decreases</u> (the backward reaction is endothermic, so this speeds up to try and reduce the temperature again).

3) The conditions that will give you <u>most ammonia</u> are <u>high pressure</u> and <u>low temperature</u>.

An equilibrium is like a particularly stubborn mule...

<u>You</u> do one thing, the <u>reaction</u> does the other. Sounds pretty <u>annoying</u>, but actually it's what gives you <u>control</u> over what happens. And in <u>industry</u>, control is what makes the whole shebang profitable.

Warm-Up and Exam Questions

Some of this stuff can be a bit tricky — have a go at these to make sure you've got it.

Warm-Up Questions

1) Is energy released when bonds are formed or when bonds are broken?
2) Which symbol is used to represent the energy change in a reaction?
3) What is the effect of a catalyst on the activation energy of a reaction?
4) What is the effect on an equilibrium of increasing the temperature if the forward reaction is endothermic?
5) What is the effect on an equilibrium of increasing the temperature if the forward reaction is exothermic?
6) For the reaction: $N_2(g) + O_2(g) \rightleftharpoons 2NO(g)$, what would be the effect on the equilibrium of changing the gas pressure?

Exam Questions

1 Charlotte investigates the effect of pressure and temperature on the chemical equilibrium:

$$A (g) + B (g) \rightleftharpoons 3C (g) \qquad \text{where A, B and C are gas molecules.}$$

She presents her findings in a graph.

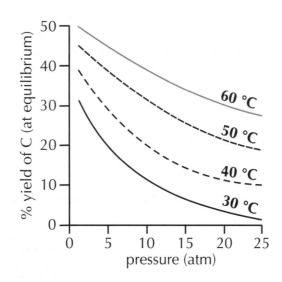

(a) At 10 atm pressure, what is the percentage yield of C at 40 °C?

(1 mark)

(b) What is the effect on the equilibrium of increasing the pressure?
Explain why this happens.

(3 marks)

(c) (i) Explain the effect on the equilibrium of increasing the temperature.

(2 marks)

(ii) Is the forward reaction endothermic or exothermic?

(1 mark)

Exam Questions

2   When methane burns in air it produces carbon dioxide and water, as shown in the diagram:

$$H-\overset{\overset{\displaystyle H}{|}}{\underset{\underset{\displaystyle H}{|}}{C}}-H \quad + \quad \begin{matrix} O=O \\ O=O \end{matrix} \quad \rightarrow \quad O=C=O \quad + \quad \begin{matrix} H-O-H \\ H-O-H \end{matrix}$$

The bond energies for each bond in the above molecules are given below.

   Bond energies (kJ/mol):   C–H 414   O=O 494   C=O 800   O–H 459

(a)   Which two types of bond are broken during the reaction?

(1 mark)

(b)   Calculate an energy value (in kJ/mol) for:

   (i)   the total bonds broken.

(1 mark)

   (ii)   the total bonds formed.

(1 mark)

   (iii) the difference between the bonds formed and the bonds broken.

(1 mark)

(c)   Use the values from part (b) to explain why the reaction is exothermic.

(1 mark)

3   The amount of energy produced by two different fuels was compared.
1 g of each fuel was burned and the heat produced was used to increase the
temperature of 100 cm³ of water. The temperature rise for fuel A was 21 °C
and for fuel B it was 32 °C. (The specific heat capacity of water is 4.2 J g⁻¹ K⁻¹.)

(a)   Why must the same volume of water be used each time?

(1 mark)

(b)   Calculate the heat energy transferred to the water from Fuel A, if the water
weighs 100 g.

(2 marks)

(c)   Complete the diagrams to compare the energy changes caused by the two fuels.

(2 marks)

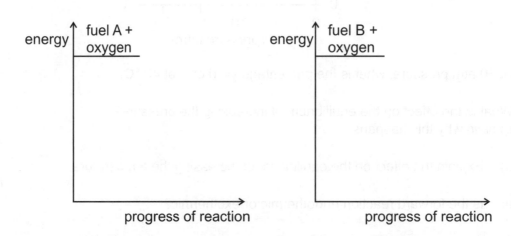

Revision Summary for Section 9

Well, I don't think that was too bad. Four things affect the rate of reactions, there are loads of ways to measure reaction rates and it's all explained by collision theory. Reactions can be endothermic or exothermic, and quite a few of them are reversible. Easy. Ahem.

Anyway, here are some more of those nice questions that you enjoy so much. If there are any you can't answer, go back to the appropriate page, take another look, then try again.

1) What are the four factors that affect the rate of a reaction?

2) Describe three different ways of measuring the rate of a reaction.

3) A student carries out an experiment to measure the effect of surface area on the reaction between marble and hydrochloric acid. He measures the amount of gas given off at regular intervals.
   a) What factors must he keep constant for it to be a fair test?

   b)* He uses four samples for his experiment:
   Sample A – 10 g of powdered marble
   Sample B – 10 g of small marble chips
   Sample C – 10 g of large marble chips
   Sample D – 5 g of powdered marble
   Sketch a typical set of graphs for this experiment.

4) Explain how higher temperature, higher concentration and larger surface area increase the frequency of successful collisions between particles.

5) What is activation energy?

6) Discuss the advantages and disadvantages of using catalysts in industrial processes.

7) What is an exothermic reaction? Give three examples.

8) The reaction to split ammonium chloride into ammonia and hydrogen chloride is endothermic. What can you say for certain about the reverse reaction?

9) An acid and an alkali were mixed in a polystyrene cup, as shown to the right. The acid and alkali were each at 20 °C before they were mixed. After they were mixed, the temperature of the solution reached 24 °C.
   a) State whether this reaction is exothermic or endothermic.
   b) Explain why the cotton wool is used.

20 cm³ of dilute sulfuric acid + 20 cm³ of dilute sodium hydroxide solution

cotton wool

10) Explain why the price of bananas might rise if we keep burning so much fuel.

11) The apparatus below is used to measure how much energy is released when pentane is burnt. It takes 4.2 joules of energy to heat 1 g of water by 1 °C.
    a)* Using the following data, and the equation $Q = mc\,\Delta T$, calculate the amount of energy per gram of pentane.

Mass of empty copper can	64 g
Mass of copper can + water	116 g

Initial temperature of water	17 °C
Final temperature of water	47 °C

Mass of spirit burner + pentane before burning	97.72 g
Mass of spirit burner + pentane after burning	97.37 g

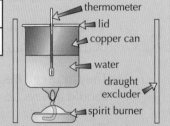

thermometer
lid
copper can
water
draught excluder
spirit burner

   b) A data book says that pentane has 49 kJ/g of energy. Why is the amount you calculated different?

12) a) Draw energy level diagrams for exothermic and endothermic reactions.
    b) Explain how bond breaking and forming relate to these diagrams.

13) What is a reversible reaction? Explain why it could reach an equilibrium.

14) Describe how three different factors affect the position of equilibrium.

15) For the reaction: $N_2(g) + 3H_2(g) \rightleftharpoons 2NH_3(g)$, what is the effect on the reaction of removing the NH_3 produced from the reaction vessel?

* Answers on page 260

Analytical Procedures

The next few pages are all about finding out exactly what is contained in a <u>mystery substance</u>.

Qualitative Analysis Tells You What a Sample Contains

1) <u>Qualitative</u> analysis tells you <u>which substances</u> are present in a <u>sample</u>.

2) It <u>doesn't tell you how much</u> of each substance there is — that's where <u>quantitative</u> analysis comes in.

Quantitative Analysis Tells You How Much it Contains

1) <u>Quantitative</u> analysis tells you <u>how much</u> of a substance is present in a sample.

2) It can be used to work out the <u>molecular formula</u> of the sample. E.g. if you had a sample containing <u>carbon</u> and <u>hydrogen</u> you'd know it was a hydrocarbon, but <u>without</u> quantitative analysis, you won't know if it's methane, butane or even 3,4-dimethylheptane...

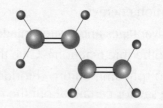

Standard Procedures Mean People Do Things the Same Way

Whether testing chemicals or measuring giraffes, scientists follow '<u>standard procedures</u>' — clear <u>instructions</u> describing <u>exactly how</u> to carry out these practical tasks.

1) Standard procedures are <u>agreed methods of working</u> — they are chosen because they're the <u>safest</u>, <u>most effective</u> and <u>most accurate methods</u> to use.

2) Standard procedures can be agreed <u>within a company</u>, <u>nationally</u>, or <u>internationally</u>.

3) They're <u>useful</u> because wherever and whenever a test is done, the result should always be the same — it should give <u>reliable</u> results every time.

4) There are <u>standard procedures</u> for the <u>collection</u> and <u>storage</u> of a sample, as well as how it should be <u>analysed</u>.

Don't get qualitative and quantitative analysis confused...

Qualitative tells you what's in a sample, <u>quant</u>itative tells you the <u>quant</u>ity. Similar words, but very different meanings. Standard procedures are a really important part of science too. If everyone was doing things differently it would all get very messy, because scientists couldn't compare their results.

Analytical Procedures

More about <u>analysing samples</u>...

Chemical Analysis *is Carried Out On* Samples

1) You usually analyse just a <u>sample</u> of the material under test. There's quite a few reasons for this.
2) It might be very difficult to test <u>all</u> of the material if you've got an awful lot of it — or you might want to just test a <u>small</u> bit so that you can <u>use</u> the rest for something else.

3) Taking a sample also means that if something goes <u>wrong</u> with the test, you can go <u>back</u> for another sample and <u>try again</u>.
4) A sample must <u>represent</u> the <u>bulk</u> of the <u>material</u> being tested — it wouldn't tell you anything very useful if it didn't.

Samples *are Analysed in* Solution

Samples are usually tested in <u>solution</u>. A solution is made by <u>dissolving</u> the sample in a <u>solvent</u>. There are <u>two</u> types of solution — <u>aqueous</u> and <u>non-aqueous</u>. Which type of solution you use depends on the type of substance you're testing.

> An <u>aqueous</u> solution means the solvent is <u>water</u>. They're shown by the <u>state symbol (aq)</u>.

> A <u>non-aqueous</u> solution means the solvent is anything <u>other than</u> water — e.g. <u>ethanol</u>.

Imagine you have a <u>mystery substance</u> — you don't know what it is, but you need to find out. If it's an <u>ionic compound</u> it'll have a positive and a negative part. So, first off, some tests for <u>positive ions</u>.

Flame Tests *Identify* Metal Ions

Compounds of some metals burn with a characteristic colour.

1) You can test for various <u>metal ions</u> by putting your substance in a <u>flame</u> and seeing what <u>colour</u> the flame goes.

Remember, metals always form positive ions. There's more about testing for positive ions on the next page.

Lithium, Li^+, gives a crimson flame.
Sodium, Na^+, gives a yellow flame.
Potassium, K^+, gives a lilac flame.
Calcium, Ca^{2+}, gives a red flame.
Barium, Ba^{2+}, gives a green flame.

2) To flame-test a compound in the lab, dip a <u>clean wire loop</u> into a sample of the compound, and put the wire loop in the clear blue part of the Bunsen flame (the hottest bit). First make sure the wire loop is really clean by dipping it into <u>hydrochloric acid</u> and rinsing it with <u>distilled water</u>.

Testing for metals is flaming useful...

To find out what your mystery ion is, start off with a <u>flame test</u>. If that doesn't give you an answer, then try the <u>sodium hydroxide</u> test (see the next page). If that doesn't work, call a detective. Or maybe not.

Tests for Positive Ions

If the flame test from the last page doesn't tell you what metal ion you're working with, then try the sodium hydroxide test described on this page. But first a bit about ionic equations....

Ionic Equations Show Just the Useful Bits of Reactions

1) The reactions in the table below are ionic equations.
   Ionic equations are 'half' a full equation, if you like. For example:

$$Ca^{2+}_{(aq)} + 2OH^-_{(aq)} \longrightarrow Ca(OH)_{2(s)}$$

2) This just shows the formation of calcium hydroxide from the calcium ions and the hydroxide ions. The full equation in the above reaction would be (if you started off with calcium chloride, say):

$$CaCl_{2(aq)} + 2NaOH_{(aq)} \longrightarrow Ca(OH)_{2(s)} + 2NaCl_{(aq)}$$

3) But the formation of sodium chloride is of no great interest here
   — it's not helping to identify the compound.

4) So the ionic equation just concentrates on the good bits.

Some Metal Ions Form a Coloured Precipitate With NaOH

This is also a test for metal ions, but it's slightly more involved. Concentrate now...

1) Many metal hydroxides are insoluble and precipitate out of solution when formed.
   Some of these hydroxides have a characteristic colour.

2) So in this test you add a few drops of sodium hydroxide solution to a solution of your mystery compound — all in the hope of forming an insoluble hydroxide.

3) If you get a coloured insoluble hydroxide you can then tell which metal was in the compound.

"Metal"	Colour of precipitate	Ionic Reaction
Calcium, Ca^{2+}	White	$Ca^{2+}_{(aq)} + 2OH^-_{(aq)} \rightarrow Ca(OH)_{2\,(s)}$
Copper(II), Cu^{2+}	Blue	$Cu^{2+}_{(aq)} + 2OH^-_{(aq)} \rightarrow Cu(OH)_{2\,(s)}$
Iron(II), Fe^{2+}	Green	$Fe^{2+}_{(aq)} + 2OH^-_{(aq)} \rightarrow Fe(OH)_{2\,(s)}$
Iron(III), Fe^{3+}	Brown	$Fe^{3+}_{(aq)} + 3OH^-_{(aq)} \rightarrow Fe(OH)_{3\,(s)}$
Aluminium, Al^{3+}	White at first. But then redissolves in excess NaOH to form a colourless solution.	$Al^{3+}_{(aq)} + 3OH^-_{(aq)} \rightarrow Al(OH)_{3\,(s)}$ then $Al(OH)_{3\,(s)} + OH^-_{(aq)} \rightarrow Al(OH)_4^-_{(aq)}$
Magnesium, Mg^{2+}	White	$Mg^{2+}_{(aq)} + 2OH^-_{(aq)} \rightarrow Mg(OH)_{2\,(s)}$

It isn't any old ion — get a positive identification...

Just think of an ionic equation as a bit like Match of the Day — an edited highlights package.

Tests for Negative Ions

So now maybe you know what the <u>positive</u> part of your mystery substance is (see pages 185 and 186). Now it's time to test for the <u>negative</u> bit.

Testing for **Carbonates** — Check for CO_2

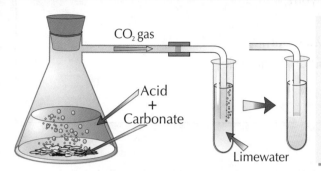

CO$_2$ gas

Acid + Carbonate

Limewater

First things first — the test for carbon dioxide (CO_2).

1) You can test to see if a gas is <u>carbon dioxide</u> by bubbling it through <u>limewater</u>. If it is <u>carbon dioxide</u>, the <u>limewater turns cloudy</u>.

2) You can use this to test for <u>carbonate</u> ions (CO_3^{2-}), since carbonates react with <u>dilute acids</u> to form <u>carbon dioxide</u>.

$$\text{acid + carbonate} \rightarrow \text{salt + water + carbon dioxide}$$

Tests for **Halides** and **Sulfates**

You can test for certain ions by seeing if a <u>precipitate</u> is formed after these reactions...

Halide ions

To test for <u>chloride</u> (Cl$^-$), <u>bromide</u> (Br$^-$) or <u>iodide</u> (I$^-$) ions, add <u>dilute nitric acid</u> (HNO$_3$), followed by <u>silver nitrate solution</u> (AgNO$_3$).

A <u>chloride</u> gives a white precipitate of <u>silver chloride</u>.

$$Ag^+_{(aq)} + Cl^-_{(aq)} \longrightarrow AgCl_{(s)}$$

A <u>bromide</u> gives a cream precipitate of <u>silver bromide</u>.

$$Ag^+_{(aq)} + Br^-_{(aq)} \longrightarrow AgBr_{(s)}$$

An <u>iodide</u> gives a yellow precipitate of <u>silver iodide</u>.

$$Ag^+_{(aq)} + I^-_{(aq)} \longrightarrow AgI_{(s)}$$

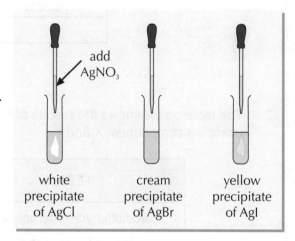

add AgNO$_3$

| white precipitate of AgCl | cream precipitate of AgBr | yellow precipitate of AgI |

Sulfate ions

1) To test for a <u>sulfate</u> ion (SO_4^{2-}), <u>add dilute HCl</u>, followed by <u>barium chloride solution</u>, BaCl$_2$.

2) A <u>white</u> precipitate of <u>barium sulfate</u> means the original compound was a sulfate.

$$Ba^{2+}_{(aq)} + SO_4^{2-}_{(aq)} \longrightarrow BaSO_{4(s)}$$

It may need loads of tests to work out what a substance is...

Well, now you know all the common tests for ions. If you ever have a sample of a mystery ionic compound (or a description of how it behaves in these tests) you should be able to work out what it is.

Warm-Up and Exam Questions

Lots of tests to remember on the last few pages. Try these questions to test your memory.

Warm-Up Questions

1) What does qualitative analysis of a sample tell you?
2) What is a non-aqueous solution?
3) What metal ion would produce a lilac flame when burnt?
4) What colour is the precipitate formed when sodium hydroxide is added to a solution of copper(II) ions?
5) How would you test for carbon dioxide?

Exam Questions

1   Kelly carried out flame tests on compounds of four different metal ions.
Complete the table below showing her results.

Flame colour	Metal ion
green	
	Li^+
yellow	
	Ca^{2+}

(4 marks)

2   The table below shows the results of a series of chemical tests conducted on two unknown compounds, X and Y.

TEST	OBSERVATION	
	COMPOUND X	COMPOUND Y
sodium hydroxide solution	white precipitate	no precipitate
hydrochloric acid & barium chloride solution	no precipitate	no precipitate
flame test	red flame	lilac flame
nitric acid & silver nitrate solution	white precipitate	yellow precipitate

(a)   What is the chemical name of compound X?

(2 marks)

(b)   What is the chemical name of compound Y?

(2 marks)

Exam Questions

3   A bottle of a chemical solution is labelled 'iron(II) sulfate'.

   (a)   Describe a chemical test to confirm that the solution contains iron(II) ions.

(2 marks)

   (b)   Describe a chemical test to confirm that the solution contains sulfate ions.

(2 marks)

4   You are provided with a solution of a halide salt. Describe how you would test this solution to identify whether the solution is of a chloride, bromide or iodide.

(4 marks)

5   William conducted a series of tests on several solutions of ionic compounds to identify the positive ions.

Complete the table below by writing the correct symbol for the positive ion that William has identified in each case.

TEST	OBSERVATION	ION
sodium hydroxide solution	white precipitate that redissolves with excess sodium hydroxide	
sodium hydroxide solution	blue precipitate	
sodium hydroxide solution and then flame test	white precipitate with sodium hydroxide and brick-red flame	

(3 marks)

6   Salim reacted hydrochloric acid with an iron(III) compound.

   (a)   During the reaction a gas formed that turned limewater cloudy.
Name the gas.

(1 mark)

   (b)   (i)   Describe a test that could be used to show that iron(III) ions were present in the compound.

(2 marks)

        (ii)   Give the ionic equation that shows the reaction of the iron(III) ions during the test.

(2 marks)

Gas Tests and Spectroscopy

It's no good carefully <u>collecting</u> a gas if you've then got <u>no idea</u> what you've collected...

There are Tests for Common Gases

Chlorine

Chlorine <u>bleaches</u> damp <u>litmus paper</u>, turning it white. (It may turn <u>red</u> for a moment first though — that's because a solution of chlorine is <u>acidic</u>.)

damp litmus paper

Oxygen

You can <u>test</u> for oxygen by checking if the gas <u>relights</u> a <u>glowing splint</u>.

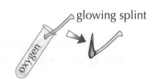

glowing splint

Hydrogen

Hydrogen makes a "<u>squeaky pop</u>" with a <u>lighted splint</u>. (The noise comes from the hydrogen burning with the oxygen in the air to form H_2O.)

Each Element Gives a Characteristic Line Spectrum

1) When <u>heated</u>, the <u>electrons</u> in an atom are <u>excited</u>, and <u>release energy as light</u>.
2) The wavelengths emitted can be <u>recorded</u> as a <u>line spectrum</u>.
3) <u>Different elements</u> emit <u>different wavelengths</u> of light. This is due to each element having a different <u>electron arrangement</u> (see page 18).
4) So each element has a <u>different pattern</u> of wavelengths, and a different line spectrum.
5) This means that line spectrums can be used to <u>identify elements</u>.
6) The practical technique used to produce line spectrums is called <u>spectroscopy</u>.

A line spectrum for an element will look something like this:

Line Spectrums Have Identified New Elements

New practical techniques (e.g. spectroscopy) have allowed scientists to <u>discover new elements</u>. Some of these elements simply wouldn't have been discovered without the development of these <u>techniques</u>.

- <u>Caesium</u> and <u>rubidium</u> were both discovered by their line spectrum.
- <u>Helium</u> was discovered in the line spectrum of the Sun.

Chromatography

Chromatography is a really good method of finding out what substances are made of.

Chromatography uses Two Phases

Chromatography is an analytical method used to separate the substances in a mixture.
You can then use it to identify the substances. There are lots of different types of
chromatography — but they all have two 'phases':

> - A mobile phase — where the molecules can move. This is always a liquid or a gas.
> - A stationary phase — where the molecules can't move. This can be a solid or a really thick liquid.

1) The components in the mixture separate out as the mobile phase moves across the stationary phase.

2) How quickly a chemical moves depends on how it "distributes" itself between the two phases
   — this is why different chemicals separate out and end up at different points (see below).

3) The molecules of each chemical constantly move between the mobile and the stationary phases.

4) They are said to reach a "dynamic equilibrium" — at equilibrium the amount leaving the
   stationary phase for the mobile phase is the same as the amount leaving the mobile phase for the
   stationary phase. But be careful, this doesn't (necessarily) mean there is the same amount
   of chemical in each phase.

*Don't panic — this will all make more
sense once you've read the rest of the page...*

In Paper Chromatography the Stationary Phase is Paper

1) In paper chromatography, a spot of the substance being tested is
   put onto a baseline on the paper.

2) The bottom of the paper is placed in a beaker containing a solvent,
   such as ethanol or water. The solvent is the mobile phase.

3) The stationary phase is the chromatography paper (often filter paper).

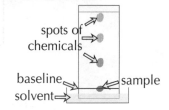

spots of
chemicals

baseline
solvent ⇒

sample

Here's what happens:

1) The solvent moves up the paper.

2) The chemicals in the sample dissolve in the solvent and move between it and the paper.
   This sets up an equilibrium between the solvent and the paper.

3) When they're in the mobile phase the chemicals move up the paper with the solvent.

4) Before the solvent reaches the top of the paper, the paper is removed from the beaker.

5) The different chemicals in the sample form separate spots on the paper.
   The chemicals that spend more time in the mobile phase than the stationary phase form
   spots further up the paper.

The amount of time the molecules spend in each phase depends on two things:

> 1) How soluble they are in the solvent.
> 2) How attracted they are to the paper.

So molecules with a higher solubility in the solvent, and which are less attracted to the paper,
will spend more time in the mobile phase — and they'll be carried further up the paper.

Chromatography

Chromatography isn't just about phases — it can involve calculations too.

Thin-Layer Chromatography has a Different Stationary Phase

1) <u>Thin-layer chromatography</u> (TLC) is very similar to paper chromatography, but the <u>stationary phase</u> is a thin layer of solid — e.g. <u>silica gel</u> spread onto a glass plate.

2) The <u>mobile phase</u> is a solvent such as ethanol (just like in paper chromatography).

You can Calculate the R_f Value for Each Chemical

1) The result of chromatography analysis is called a <u>chromatogram</u>.

2) Some of the spots on the chromatogram might be <u>colourless</u>. If they are, you need to use a <u>locating agent</u> to show where they are, e.g. you might have to spray the chromatogram with a reagent.

3) You can work out the <u>R_f values</u> for <u>spots</u> (solutes) on a chromatogram. An R_f value is the <u>ratio</u> between the distance travelled by the dissolved substance (the solute) and the distance travelled by the solvent. You can find them using the formula:

$$R_f = \frac{\text{distance travelled by solute}}{\text{distance travelled by solvent}}$$

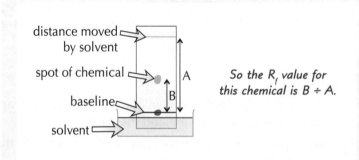

So the R_f value for this chemical is B ÷ A.

4) Chromatography is often carried out to see if a certain substance is present in a mixture. You run a <u>pure, known sample</u> of the substance alongside the unknown mixture. If the R_f values match, the substances may be the <u>same</u> (although it doesn't definitely prove they are the same).

5) Chemists use substances called <u>standard reference materials</u> (SRMs) to check the identities of substances. These have carefully <u>controlled concentrations and purities</u>.

Learning about this — it's just a phase you go through...

Chromatography works by showing how mystery chemicals get <u>distributed</u> between mobile and stationary phases — that's what the <u>R_f value</u> represents. All chemicals get distributed differently, so that's how you can tell which is which. It's great — all you need is some paper and a bit of solvent.

Chromatography

You can do chromatography on gases too...

Gas Chromatography is a Bit More High-Tech

Gas chromatography (GC) is used to analyse unknown substances too.
If they're not already gases, then they have to be vaporised.

> * The mobile phase is an unreactive gas such as nitrogen.
> * The stationary phase is a viscous (thick) liquid, such as an oil.

The process is quite different from paper chromatography and TLC:

1) The unknown mixture is injected into a long tube coated on the inside with the stationary phase.

2) The mixture moves along the tube with the mobile phase until it comes out the other end.
   Like in the other chromatography methods, the substances are distributed between the phases.

3) The time it takes a chemical to travel through the tube is called the retention time.

4) The retention time is different for each chemical — it's what's used to identify it.

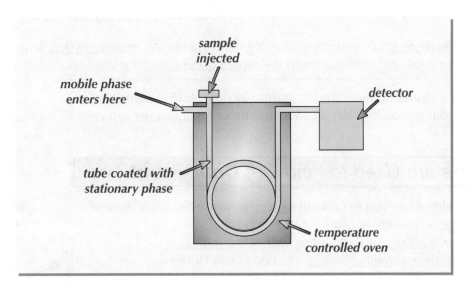

The chromatogram from gas chromatography is a graph.
Each peak on the graph represents a different chemical.

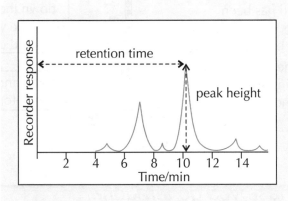

* The distance along the x-axis is the retention time — which can be looked up to find out what the chemical is.
* The peak height shows you how much of that chemical was in the sample.

Moles and Titration

This page is full of useful stuff about <u>moles</u> and <u>concentration</u>. There's also a nice little experiment you can do to find out how much alkali you need to <u>neutralise</u> an acid.

"THE MOLE" is Simply the Name Given to a Certain Number

> Just like "<u>a million</u>" is this many: 1 000 000; or "<u>a billion</u>" is this many: 1 000 000 000, "<u>a mole</u>" is this many: 602 300 000 000 000 000 000 000   or 6.023×10^{23}.

1) And that's all it is. <u>Just a number</u>. But why is it such a long number with a six at the front?

2) The answer is that when you get <u>precisely that number</u> of atoms of <u>carbon-12</u> it weighs exactly <u>12 g</u>. So, get that number of atoms or molecules, <u>of any element or compound</u>, and conveniently, they <u>weigh</u> exactly the same number of <u>grams</u> as the relative atomic mass, A_r (or M_r) of the element or compound. This is arranged <u>on purpose</u> of course, to make things easier.

3) So, you can use <u>moles</u> as a <u>unit</u> of measurement when you're talking about an amount of a substance.

Concentration is a Measure of How Crowded Things Are

The <u>concentration</u> of a solution can be measured in <u>moles per dm³</u> (i.e. <u>moles per litre</u>).
So 1 mole of stuff in 1 dm³ of solution has a concentration of <u>1 mole per dm³</u> (or 1 mol/dm³).

> The <u>more solute</u> you dissolve in a given volume, the <u>more crowded</u> the solute molecules are and the <u>more concentrated</u> the solution.

	1 litre
=	1000 cm³
=	1 dm³

Concentration can also be measured in <u>grams per dm³</u>. So 56 grams of stuff dissolved in 1 dm³ of solution has a concentration of <u>56 grams per dm³</u>.

Titrations are Used to Find Out Concentrations

You can also do titrations the other way round — adding alkali to acid.

1) Titrations also allow you to find out <u>exactly</u> how much acid is needed to <u>neutralise</u> a quantity of alkali (or vice versa).

2) You put some <u>alkali</u> in a flask, along with some <u>indicator</u> — phenolphthalein or methyl orange. You don't use Universal indicator as it changes colour gradually — and you want a <u>definite</u> colour change.

3) Add the <u>acid</u>, a bit at a time, to the alkali using a <u>burette</u> — giving the flask a regular <u>swirl</u>. Go especially <u>slowly</u> (a drop at a time) when you think the alkali's almost neutralised.

4) The indicator <u>changes colour</u> when <u>all</u> the alkali has been <u>neutralised</u> — phenolphthalein is <u>pink</u> in <u>alkalis</u> but <u>colourless</u> in <u>acids</u>, and methyl orange is <u>yellow</u> in <u>alkalis</u> but <u>red</u> in <u>acids</u>.

5) <u>Record</u> the amount of acid used to <u>neutralise</u> the alkali. It's best to <u>repeat</u> this process a few times, making sure you get (pretty much) the same answer each time.

6) You can then take the <u>mean</u> of your results.

burette containing acid

These marks down the side show the volume of acid used.

alkali and indicator

Repeating the experiment is really important

Repeating titrations and other experiments a few times helps to make sure your results are <u>reliable</u>. If you get the same result a number of times, you can have more faith in it than if it's a one-off.

Titration Calculations

The results of a titration experiment can be used to calculate concentrations of acid or alkalis...

You Can **Calculate Concentration** in **Moles** or in **Grams**

Example 1: If You Want to Know the Concentration in **MOLES** per dm^3

Say you start off with 25 cm^3 of sodium hydroxide in your flask, and you know that its concentration is 0.1 moles per dm^3.

You then find from your titration that it takes 30 cm^3 of sulfuric acid (whose concentration you don't know) to neutralise the sodium hydroxide.

You can work out the concentration of the acid in moles per dm^3.

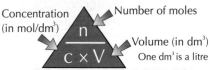

Concentration = moles ÷ volume, so you can make a handy formula triangle.

Concentration (in mol/dm^3) — $\dfrac{n}{c \times V}$ — Number of moles

Volume (in dm^3) One dm^3 is a litre

Cover up the thing you're trying to find — then what's left is the formula you need to use.

Step 1: Work out how many moles of the "known" substance you have using this formula:

Number of moles = concentration × volume
= 0.1 mol/dm^3 × (25 / 1000) dm^3
= 0.0025 moles of NaOH   *Remember: 1000 cm^3 = 1 dm^3*

Step 2: Write down the balanced equation of the reaction...

$$2NaOH + H_2SO_4 \longrightarrow Na_2SO_4 + 2H_2O$$

...and work out how many moles of the "unknown" stuff you must have had.
Using the equation, you can see that for every two moles of sodium hydroxide you had...
...there was just one mole of sulfuric acid.
So if you had 0.0025 moles of sodium hydroxide...
...you must have had 0.0025 ÷ 2 = 0.00125 moles of sulfuric acid.

Step 3: Work out the concentration of the "unknown" stuff.   *Don't forget to put the units.*
Concentration = number of moles ÷ volume
= 0.00125 mol ÷ (30 / 1000) dm^3 = 0.041666... mol/dm^3 = 0.0417 mol/dm^3

Example 2: If You Want to Know the Concentration in **GRAMS** per dm^3

You might want to find out the acid concentration in grams per cubic decimetre (grams per litre).
Don't panic — you just need another formula triangle.

Step 1: Work out the relative formula mass for the acid
(you should be given the relative atomic masses,
e.g. H = 1, S = 32, O = 16):
So, H_2SO_4 = (1 × 2) + 32 + (16 × 4) = 98

Number of moles = mass ÷ relative formula mass.

Mass (in grams)
Number of moles — $\dfrac{m}{n \times M_r}$ — Relative formula mass

Step 2: Convert the concentration in moles (that you've already worked out) into concentration in grams. So, in 1 dm^3:

Mass in grams = moles × relative formula mass
= 0.041666... × 98 = 4.08333... g   *use non-rounded answers in working*
So the concentration in g/dm^3 = 4.08 g/dm^3

Purity

It's all very well making a product, but how do you know if it's any good or not...

Some Products Need to be *Very Pure*

1) The purity of a product will improve as it's being isolated. But to get a really pure product earlier stages (such as filtration, evaporation and crystallisation) will often need to be repeated.

2) Purifying and measuring the purity of a product are particularly important steps in the chemical industry. For example, in the production of...

- Pharmaceuticals — it's really important to ensure drugs intended for human consumption are free from impurities — these could do more harm than good.
- Petrochemicals — if there are impurities or contaminants in petrol products they could cause damage to a car's engine.

Titrations Can be Used to *Measure* the *Purity* of a Substance

The purity of a compound can be calculated using a titration:

Example — Determining the Purity of Aspirin

Say you start off with 0.2 g of impure aspirin dissolved in 25 cm^3 of ethanol in your conical flask. You find from your titration that it takes 9.5 cm^3 of 4 g/dm^3 NaOH to neutralise the aspirin solution.

STEP 1: Work out the concentration of the aspirin solution.
Here's the formula — just stick the numbers in the right places.

$$\text{conc. of aspirin solution} = 4.5 \times \frac{\text{conc. of NaOH} \times \text{vol. of NaOH}}{\text{vol. of aspirin solution}}$$

This number will change depending on which acid and alkali are used.

$$= 4.5 \times \frac{4 \times (9.5 / 1000)}{25 / 1000} = \underline{6.84 \text{ g/dm}^3}$$

The volumes need to be in dm^3, not cm^3, so you need to divide by 1000 to convert it.

$$1 \text{ dm}^3 = 1000 \text{ cm}^3$$

STEP 2: Work out the mass of the aspirin. You'll need to use another formula:

$$\text{mass} = \text{concentration} \times \text{volume}$$

$$= 6.84 \times 25 / 1000 = \underline{0.171 \text{ g}}$$

This is the concentration of the aspirin solution.

In the 0.2 g that you started with, 0.171 g was aspirin.

STEP 3: Calculate the purity, using the formula:

$$\% \text{ purity} = \frac{\text{calculated mass of substance}}{\text{mass of impure substance at start}} \times 100$$

$$= 0.171 \div 0.2 \times 100 = \underline{85.5\%}$$

This is the percentage purity of the aspirin.

This is the mass of aspirin that reacted in the titration.

This is the mass of impure aspirin you started with.

Warm-Up and Exam Questions

Have a go at these questions to hone your killer exam skills. First a (fairly) gentle warm-up, and then some more exam-like questions... enjoy.

Warm-Up Questions

1) What is the mobile phase in gas chromatography?
2) What is the mass of one mole of oxygen gas?
3) Briefly describe how you'd carry out an acid-base titration.
4) How many moles of hydrochloric acid are there in 25 cm^3 of a 0.1 mol/dm^3 solution?
5) A solution of sodium carbonate, Na_2CO_3, has a concentration of 0.025 mol/dm^3.
   What is the concentration of this solution in g/dm^3?

Exam Questions

1   Scientists analysed the composition of six food colourings using chromatography.
    Four of the colourings were unknown (**1 – 4**), and the other two were known, sunrise yellow and sunset red. The results are shown below.

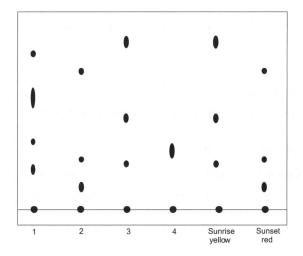

1   2   3   4   Sunrise   Sunset
                   yellow   red

(a) Which food colouring is most likely to be a pure compound?

(1 mark)

(b) Which food colouring contains at least four different compounds?

(1 mark)

(c) Which food colouring contains the same compounds as sunrise yellow?

(1 mark)

(d) Which food colouring contains the same compounds as sunset red?

(1 mark)

2   The elements caesium and rubidium were both discovered by their line spectrums.

(a) Explain how a line spectrum is formed.

(3 marks)

(b) Explain why caesium and rubidium have different line spectrums.

(3 marks)

(c) Name the technique used by scientists to produce line spectrums.

(1 mark)

Exam Questions

3   In a titration, 30.3 cm³ of a solution of 1.00 mol/dm³ sodium hydroxide was required to neutralise 25.0 cm³ of a solution of sulfuric acid.

(a)   Calculate the number of moles of sodium hydroxide used in the titration.

(2 marks)

(b)   Work out the number of moles of sulfuric acid used in the titration.

The equation is: $2NaOH$ (aq) + H_2SO_4 (aq) $\rightarrow$ Na_2SO_4 (aq) + $2H_2O$ (l)

(1 mark)

(c)   Calculate the concentration of the sulfuric acid solution.

(2 marks)

4   Jonah is concerned about the amount of acid in soft drinks. He decides to use a titration method to find the acid content of his favourite lemonade. He uses a solution of 0.1 mol/dm³ sodium hydroxide in titrations with 25 cm³ samples of the lemonade. His results are shown in the table.

	Initial burette reading (cm³)	Final burette reading (cm³)	Vol. of NaOH needed (cm³)
1	0.0	9.4	9.4
2	9.4	18.4	9.0
3	18.4	27.4	9.0

Jonah calculates that the average volume of 0.1 mol/dm³ NaOH needed is 9.0 cm³.

(a)   The first titration value was not included in calculating the average. Why not?

(1 mark)

(b)   The equation for the reaction in the titration can be written:

$HA + NaOH \rightarrow NaA + H_2O$, where HA is the acid present in the lemonade. Calculate the concentration of acid HA present in the lemonade.

(3 marks)

5   A laboratory carries out tests on a pharmaceutical drug to test its purity.

(a)   Suggest why the purity of the drug is important.

(1 mark)

(b)   One 0.23 g tablet of the drug is dissolved in 25 cm³ of ethanol. Titration with alkali shows the concentration of the drug in the solution to be 7.38 g/dm³. What is the purity of the tablet?

(2 marks)

6   Emma conducts some tests on three different gases.
Copy and complete the table below, showing her results.

Test result	Gas
Turns litmus paper white	
	Hydrogen
Relights a glowing splint	

(3 marks)

Electrolysis

Electrolysis is handy technique used to <u>break down ionic substances</u>.

Electrolysis Means "Splitting Up With Electricity"

1) If you pass an <u>electric current</u> through an <u>ionic substance</u> that's <u>molten</u> or in <u>solution</u>, it breaks down into the <u>elements</u> it's made of. This is called <u>electrolysis</u>.

2) It requires a <u>liquid</u> to <u>conduct</u> the <u>electricity</u>, called the <u>electrolyte</u>.

3) Electrolytes contain <u>free ions</u> — they're usually the <u>molten</u> or <u>dissolved ionic substance</u>.

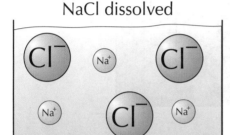

NaCl dissolved

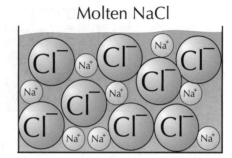

Molten NaCl

4) In either case it's the <u>free ions</u> which <u>conduct</u> the electricity and allow the whole thing to work.

5) For an electrical circuit to be complete, there's got to be a <u>flow of electrons</u>. <u>Electrons</u> are taken <u>away from</u> ions at the <u>positive electrode</u> and <u>given to</u> other ions at the <u>negative electrode</u>. As ions gain or lose electrons they become atoms or molecules and are released.

*Electrolysis Reactions Involve **Oxidation** and **Reduction***

1) <u>Oxidation</u> is a <u>gain of oxygen</u> or a <u>loss of electrons</u> (see page 154).

2) <u>Reduction</u> is a <u>loss of oxygen</u> or a <u>gain of electrons</u>.

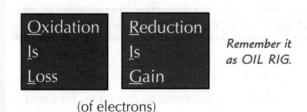

Remember it as OIL RIG.

(of electrons)

3) Electrolysis <u>ALWAYS</u> involves an oxidation and a reduction.

Electrolysis needs a liquid to conduct the electricity

Before you electrolyse a substance it has to be <u>liquid</u>. This allows the ions to <u>move</u> towards the <u>positive</u> or <u>negative</u> electrode. The next few pages are about the electrolysis of different substances.

Electrolysis of Lead Bromide

Molten lead bromide can be broken down by electrolysis. You end up with <u>lead</u> and <u>bromine</u>.

The *Electrolysis* of Molten *Lead Bromide*

When a salt (e.g. lead bromide, $PbBr_2$) is molten it will conduct electricity.

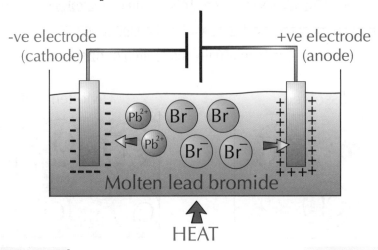

-ve electrode
(cathode)

+ve electrode
(anode)

Molten lead bromide

HEAT

<u>+ve ions</u> are attracted
to the <u>–ve electrode</u>.
Here they <u>gain</u>
<u>electrons</u> (reduction).

<u>Lead</u> is produced
at the <u>–ve electrode</u>.

<u>–ve ions</u> are attracted
to the <u>+ve electrode</u>.
Here they <u>lose</u>
<u>electrons</u> (oxidation).

<u>Bromine</u> is produced
at the <u>+ve electrode</u>.

1) At the <u>–ve electrode</u>, one lead ion <u>accepts</u> two electrons to become <u>one lead atom</u>.

2) At the <u>+ve electrode</u>, two bromide ions <u>lose</u> one electron each
    and become <u>one bromine molecule</u>.

Reactivity Affects the *Products* Formed By *Electrolysis*

1) Sometimes there are <u>more than two free ions</u> in the electrolyte.
    For example, if a salt is <u>dissolved in water</u> there will also be some H^+ and OH^- <u>ions</u>.

2) At the <u>negative electrode</u>, if <u>metal ions</u> and <u>H^+ ions</u> are present, the metal ions will <u>stay in solution</u>
    if the metal is <u>more reactive</u> than hydrogen. This is because the more reactive an element, the
    keener it is to stay as ions. So, <u>hydrogen</u> will be produced unless the metal is <u>less reactive</u> than it.

3) At the <u>positive electrode</u>, if <u>OH^-</u> and <u>halide ions</u> (Cl^-, Br^-, I^-) are present then molecules of chlorine,
    bromine or iodine will be formed. If <u>no halide</u> is present, then <u>oxygen</u> will be formed.

So, lead bromide splits into lead and bromine — I know, it's tricky

The electrolysis of lead bromide is pretty simple — <u>bromine ions</u> are <u>oxidised</u> so bromine is produced
at the positive electrode and <u>lead ions</u> are <u>reduced</u> so lead is produced at the negative electrode.
But when your substance is <u>dissolved in water</u> things get trickier — as you'll see on the next page.

Electrolysis of Sodium Chloride Solution

Time for more electrolysis. This time it's the electrolysis of salt (sodium chloride) solution.

The **Electrolysis** of **Sodium Chloride Solution**

When common salt (sodium chloride) is dissolved in water and electrolysed,
it produces three useful products — <u>hydrogen</u>, <u>chlorine</u> and <u>sodium hydroxide</u>.

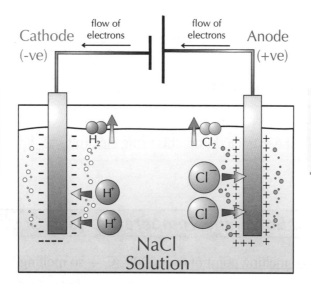

H⁺ ions are released from the water.

+ve ions are attracted to the –ve electrode. Here they <u>gain</u> <u>electrons</u> (reduction).

<u>Hydrogen</u> is produced at the <u>–ve electrode</u>.

–ve ions are attracted to the +ve electrode. Here they <u>lose</u> <u>electrons</u> (oxidation).

<u>Chlorine</u> is produced at the <u>+ve electrode</u>.

1) At the <u>negative electrode</u>, two hydrogen ions accept two electrons to become <u>one hydrogen molecule</u>.

2) At the <u>positive electrode</u>, two chloride (Cl^-) ions lose their electrons and become <u>one chlorine molecule</u>.

3) The <u>sodium ions</u> stay in solution because they're <u>more reactive</u> than hydrogen. <u>Hydroxide ions</u> from water are also left behind. This means that <u>sodium hydroxide</u> (NaOH) is left in the solution.

The **Half-Equations** — *Make Sure the Electrons Balance*

1) Half equations show the reactions at the electrodes. The main thing is to make sure the <u>number of electrons</u> is the <u>same</u> for <u>both half-equations</u>.

2) For the electrolysis of sodium chloride the half-equations are:

You need to make sure the atoms are balanced too.

Negative Electrode: $2H^+ + 2e^- \rightarrow H_2$
Positive Electrode: $2Cl^- \rightarrow Cl_2 + 2e^-$
or $2Cl^- - 2e^- \rightarrow Cl_2$

For the electrolysis of molten lead bromide (previous page) the half equations would be:
$$Pb^{2+} + 2e^- \rightarrow Pb$$
and $2Br^- \rightarrow Br_2 + 2e^-$

Products of **NaCl Electrolysis** are **Used in Industry**

The products of the electrolysis of sodium chloride solution are pretty useful in <u>industry</u>.

1) Chlorine has many uses, e.g. in the production of <u>bleach</u> and <u>plastics</u>.

2) Sodium hydroxide is a very strong <u>alkali</u> and is used <u>widely</u> in the <u>chemical industry</u>, e.g. to make <u>soap</u>.

Electrolysis of Aluminium

OK — one more example of electrolysis. This is the electrolysis of <u>aluminium oxide</u>.

*Electrolysis is Used to Remove **Aluminium** From its **Ore***

1) Aluminium's a very <u>abundant</u> metal, but it is always found naturally in <u>compounds</u>.

2) Its main ore is <u>bauxite</u>, and after mining and purifying, a <u>white powder</u> is left.

3) This is <u>pure</u> aluminium oxide, Al_2O_3.

4) The <u>aluminium</u> has to be extracted from this using <u>electrolysis</u>.

***Cryolite** is Used to **Lower** the **Temperature** (and Costs)*

1) Al_2O_3 has a very <u>high melting point</u> of over <u>2000 °C</u> — so melting it would be very <u>expensive</u>.

2) <u>Instead</u> the aluminium oxide is <u>dissolved</u> in <u>molten cryolite</u> (a less common ore of aluminium).

3) This brings the <u>temperature down</u> to about <u>900 °C</u>, which makes it much <u>cheaper</u> and <u>easier</u>.

4) The <u>electrodes</u> are made of <u>carbon</u> (graphite), a good conductor of electricity (see page 120).

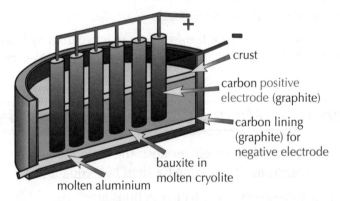

crust

carbon positive electrode (graphite)

carbon lining (graphite) for negative electrode

bauxite in molten cryolite

molten aluminium

5) <u>Aluminium</u> forms at the <u>negative electrode (cathode)</u> and <u>oxygen</u> forms at the <u>positive electrode (anode)</u>.

> Negative Electrode: $Al^{3+} + 3e^- \rightarrow Al$　　　Positive Electrode: $2O^{2-} \rightarrow O_2 + 4e^-$

6) The <u>oxygen</u> then reacts with the <u>carbon</u> in the electrode to produce <u>carbon dioxide</u>. This means that the <u>positive electrodes</u> gradually get 'eaten away' and have to be <u>replaced</u> every now and again.

It's all about lowering the cost...

The electrolysis of aluminium oxide may look <u>a bit different</u> to the examples on the other pages but don't be fooled. It's the same story — the positive aluminium ions go to the <u>negative</u> electrode and the negative oxygen ions are attracted to the <u>positive</u> electrode.

Electroplating

Electroplating coats one metal onto the surface of another. It's really useful...

Electroplating Uses Electrolysis

1) Electroplating uses electrolysis to <u>coat</u> the <u>surface of one metal</u> with <u>another metal</u>, e.g. you might want to electroplate silver onto a brass cup to make it look nice.

2) The <u>negative electrode (cathode)</u> is the <u>metal object</u> you want to plate and the <u>positive electrode (anode)</u> is the <u>pure metal</u> you want it to be plated with. You also need the <u>electrolyte</u> to contain <u>ions</u> of the <u>plating metal</u>. (The ions that plate the metal object come from the solution, while the positive electrode keeps the solution 'topped up'.)

<u>Example</u>: To electroplate <u>silver</u> onto a <u>brass cup</u>, you'd make the <u>brass cup</u> the negative electrode (to attract the positive silver ions), a lump of <u>pure silver</u> the positive electrode and dip them in a solution of <u>silver ions</u>, e.g. silver nitrate.

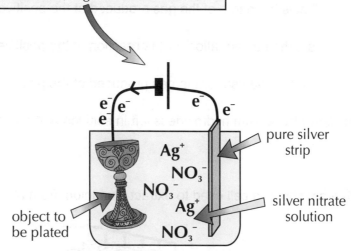

object to be plated

pure silver strip

silver nitrate solution

3) There are lots of different <u>uses</u> for electroplating:

- <u>Decoration</u>: <u>Silver</u> is <u>attractive</u>, but very <u>expensive</u>. It's much <u>cheaper</u> to plate a brass cup with silver, than it is to make the cup out of solid silver — but it looks just as <u>pretty</u>.
- <u>Conduction</u>: Metals like <u>copper</u> conduct <u>electricity</u> well — because of this they're often used to plate metals for <u>electronic circuits</u> and <u>computers</u>.

Silver electroplated text is worth a fortune...

There are loads of metals you can use for electroplating — silver and copper are two examples, but they're not the only ones. The tricky bit is remembering that the metal <u>object you want to plate</u> is the <u>negative electrode</u> and the <u>metal</u> you're plating it with is the <u>positive electrode</u>.

Warm-Up and Exam Questions

It's question time again. You know the drill. Off you go...

Warm-Up Questions

1) What state must an ionic compound be in if it's to be used as an electrolyte?
2) In electrolysis, what is meant by the terms oxidation and reduction?
3) At which electrode are metals deposited during electrolysis?
4) During the electrolysis of molten lead bromide, which gas is produced at the positive electrode?
5) An object requires electroplating. Which electrode should it be used as?

Exam Questions

1   When sodium chloride solution is electrolysed a gas is produced at each electrode.
   (a) (i)   What is the name of the gas produced at the negative electrode?

(1 mark)

      (ii) State the half equation for the reaction at the negative electrode.

(1 mark)

   (b) (i)   What is the name of the gas produced at the positive electrode?

(1 mark)

      (ii) State the half equation for the reaction at the positive electrode.

(1 mark)

      (iii) Suggest one use for the gas produced at the positive electrode.

(1 mark)

   (c) Explain why sodium hydroxide is left in solution at the end of the reaction.

(3 marks)

2   The diagram shows a cell used to extract aluminium from aluminium oxide.

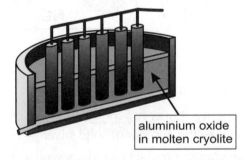

aluminium oxide
in molten cryolite

   (a) Explain why the aluminium oxide is dissolved in molten cryolite.

(2 marks)

   (b) Complete the half-equation below for the reaction at the negative electrode.

   _____ + 3e⁻ → _____

(1 mark)

   (c) The positive electrode is made out of carbon. Explain why it will need to be replaced over time.

(3 marks)

Exam Questions

3 Electroplating is used to coat one metal with a thin layer of another metal.

   (a) Bruce wants to coat a brass cup with silver using electroplating.
       He sets up the apparatus shown below.

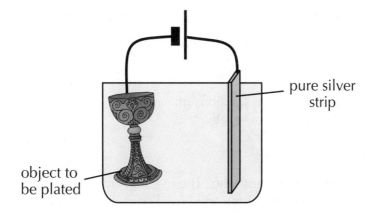

      (i)   What acts as the anode during the electroplating process?

(1 mark)

      (ii) Describe the electrolyte Bruce will need to use. Explain your answer.

(2 marks)

      (iii) Give **one** reason why Bruce might want to plate the cup with silver.

(1 mark)

   (b) Copper is often plated onto metals used in electronic circuits.
       Give the property of copper that makes it ideal for this purpose.

(1 mark)

4 Lead bromide can be broken down into its elements by electrolysis using the apparatus
   shown in the diagram below.

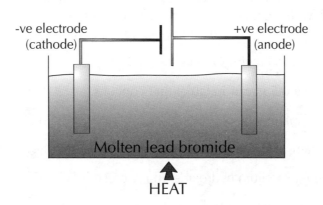

   (a) Explain why the lead bromide must be molten.

(2 marks)

   (b) Lead bromide is an ionic substance. Describe what happens to the ions
       in lead bromide, in terms of electron loss and gain, during electrolysis.

(2 marks)

   (c) Write the half equations for the reactions that take place at the anode and cathode
       during the electrolysis of molten lead bromide.

(2 marks)

Revision Summary for Section 10

And that's it... the end of another section. Which means it's time for some more questions. There's no point in trying to duck out of these — they're the best way of testing that you've learned everything in this topic. For these questions you'll need to know some common chemical tests. If you find you don't know them, look back in the book.

1) What is the difference between qualitative and quantitative analysis?

2) What is meant by the phrase 'standard procedures'? Why are they important?

3) Describe two ways of testing for metal ions.

4) How would you distinguish between solutions of:
    a) magnesium sulfate and aluminium sulfate,
    b) sodium bromide and sodium iodide,
    c) copper nitrate and copper sulfate?

5)* You are conducting a reaction in a test tube. The reaction gives off a gas.
    You stick a lighted splint into the test tube and hear a squeaky pop. What does this tell you?

6) What can line spectrums be used for?

7) Describe the two phases in chromatography.

8) What are the mobile and stationary phases in paper chromatography and thin-layer chromatography?

9)* What is the R_f value of a chemical that moves 4.5 cm when the solvent moves 12 cm?

10) What is the stationary phase in gas chromatography?

11) What is meant by 'retention time'?

12) What does the height of a peak on a gas chromatogram show?

13) Name a suitable indicator you could use in the titration of sulfuric acid and sodium hydroxide.

14)*In a titration, 49 cm³ of hydrochloric acid was required to neutralise 25 cm³
    of sodium hydroxide with a concentration of 0.2 moles per dm³.
    Calculate the concentration of the hydrochloric acid in: a) mol/dm³   b) g/dm³

15) Why is purification of a product important?

16)*Calculate the purity of a 0.5 g aspirin tablet that contains 0.479 g of aspirin.

17) What is electrolysis? Explain why only liquids can be electrolysed.

18) Draw a detailed diagram with half equations showing the electrolysis of sodium chloride.

19) Give one industrial use of sodium hydroxide and two uses of chlorine.

20) In the electrolysis of aluminium oxide name the substance that forms at:
    (a) the anode (the positive electrode).
    (b) the cathode (the negative electrode).

21) Give two different uses of electroplating.

* Answers on page 262

Hardness of Water

Water where you live might be <u>hard</u> or <u>soft</u>. It depends on the <u>rocks</u> your water meets on its way to you.

Hard Water Makes Scum and Scale

1) With <u>soft water</u>, you get a nice <u>lather</u> with soap. But with <u>hard water</u> you get a <u>nasty scum</u> instead — unless you're using a soapless detergent. The problem is dissolved <u>calcium ions</u> and <u>magnesium ions</u> in the water (see below) reacting with the soap to make <u>scum</u> which is insoluble. So to get a decent lather you need to use <u>more soap</u> — and because soap <u>isn't free</u>, that means <u>more money</u> going down the drain.

2) When <u>heated</u>, hard water also forms furring or <u>scale</u> (mostly calcium carbonate) on the insides of pipes, boilers and kettles. Badly scaled-up pipes and boilers reduce the <u>efficiency</u> of heating systems, and may need to be <u>replaced</u> — all of which costs money. Scale can even <u>eventually block pipes</u>.

3) <u>Scale</u> is also a bit of a <u>thermal insulator</u>. This means that a <u>kettle</u> with scale on the <u>heating element</u> takes <u>longer to boil</u> than a <u>clean</u> non-scaled-up kettle — so it becomes <u>less efficient</u>.

Hardness is Caused by Ca^{2+} and Mg^{2+} Ions

1) Most hard water is hard because it contains lots of <u>calcium ions</u> and <u>magnesium ions</u>.

2) Rain falling on some types of rocks (e.g. <u>limestone</u>, <u>chalk</u> and <u>gypsum</u>) can dissolve compounds like <u>magnesium sulfate</u> (which is soluble), and <u>calcium sulfate</u> (which is also soluble, though only a bit).

Hard Water Isn't All Bad

1) Ca^{2+} ions are good for healthy <u>teeth</u> and <u>bones</u>.

2) Studies have found that people who live in <u>hard water</u> areas are at <u>less risk</u> of developing <u>heart disease</u> than people who live in soft water areas. This could be to do with the <u>minerals</u> in hard water.

Hardness of Water

Remove Dissolved Ca²⁺ and Mg²⁺ Ions to Make Hard Water Soft

There are two kinds of hardness — temporary and permanent.
Temporary hardness is caused by the hydrogencarbonate ion, HCO_3^-, in $Ca(HCO_3)_2$.
Permanent hardness is caused by dissolved calcium sulfate (among other things).

1) Temporary hardness is removed by boiling. When heated, the calcium hydrogencarbonate decomposes to form calcium carbonate which is insoluble. This solid is the 'limescale' on your kettle.

   e.g.

   $$\text{calcium hydrogencarbonate} \rightarrow \text{calcium carbonate} + \text{water} + \text{carbon dioxide}$$
   $$Ca(HCO_3)_{2(aq)} \rightarrow CaCO_{3(s)} + H_2O_{(l)} + CO_{2(g)}$$

   This won't work for permanent hardness, though. Heating a sulfate ion does nowt.

2) Both types of hardness can be softened by adding washing soda (sodium carbonate, Na_2CO_3) to it. The added carbonate ions react with the Ca^{2+} and Mg^{2+} ions to make an insoluble precipitate of calcium carbonate and magnesium carbonate. The Ca^{2+} and Mg^{2+} ions are no longer dissolved in the water so they can't make it hard.

   e.g. $$Ca^{2+}_{(aq)} + CO_3^{2-}_{(aq)} \rightarrow CaCO_{3(s)}$$

3) Both types of hardness can also be removed by running water through 'ion exchange columns' which are sold in shops. The columns have lots of sodium ions (or hydrogen ions) and 'exchange' them for calcium or magnesium ions in the water that runs through them.

   e.g. $$Na_2Resin_{(s)} + Ca^{2+}_{(aq)} \rightarrow CaResin_{(s)} + 2Na^+_{(aq)}$$

   ('Resin' is a huge insoluble resin molecule.)

And if the water's really hard, you can chip your teeth...

Hard water — good thing or bad thing... Well, it provides minerals that are good for health, but it creates an awful lot of unnecessary expense. In hard water areas, you need more soap to get a lather, it takes longer (and therefore more electricity) to boil water (as heating elements get furred up), and you need to get your pipes replaced more often. It's a bit of a drag.

Hardness of Water

With soft water, you get a nice <u>lather</u> with soap. Not so with hard water...

Use Titration to **Compare the Hardness** of **Water Samples**

Method

1) Fill a burette with <u>50 cm³ of soap solution</u>.

2) Add <u>50 cm³</u> of the first <u>water</u> sample into a flask.

3) Use the burette to add <u>1 cm³ of soap solution</u> to the flask.

4) Put a <u>bung</u> in the flask and <u>shake</u> for 10 seconds.

5) <u>Repeat</u> steps 3 and 4 until a <u>good lasting lather</u> is formed. (A lasting lather is one where the <u>bubbles cover the surface</u> for <u>at least 30 seconds</u>.)

6) <u>Record</u> how much soap was needed to create a lasting lather.

7) <u>Repeat</u> steps 1-6 with the other water samples.

8) Next, <u>boil fresh samples</u> of each type of water for <u>ten minutes</u>, and <u>repeat</u> the experiment.

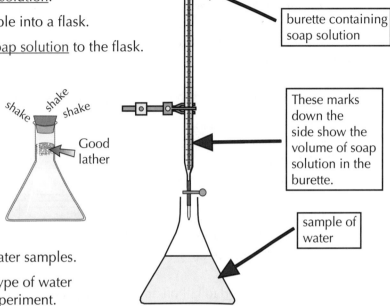

burette containing soap solution

These marks down the side show the volume of soap solution in the burette.

sample of water

shake shake shake

Good lather

Results

This method was carried out on <u>3 different samples of water</u> — <u>distilled</u> water, <u>local tap water</u> and <u>imported tap water</u>. Here's the <u>table of results</u>:

Sample	Volume of soap solution needed to give a good lather	
	using unboiled water in cm³	using boiled water in cm³
Distilled	1	1
Local water	7	1
Imported water	14	8

The results tell you the following things about the water:

1) Distilled water contains little or no <u>hardness</u> — only the <u>minimum</u> amount of soap was needed.

2) The sample of <u>imported water</u> contains <u>more hardness</u> than <u>local water</u> — <u>more soap</u> was needed to produce a lather.

3) The local water contains only <u>temporary hardness</u> — all the hardness is <u>removed by boiling</u>. You can tell because the same amount of soap was needed for <u>boiled local water</u> as for <u>distilled water</u>.

4) The imported water contains both <u>temporary</u> and <u>permanent hardness</u>. 8 cm³ of soap is still needed to produce a lather after boiling.

5) If your brain's really switched on, you'll see that the local water and the imported water contain the <u>same amount</u> of <u>temporary hardness</u>. In both cases, the amount of soap needed in the <u>boiled</u> sample is <u>6 cm³ less</u> than in the <u>unboiled</u> sample.

Water Quality

It's easy to take water for granted... turn on the tap, and there it is — nice, clean water. The water you drink's been round the block a few times — so there's some <u>fancy chemistry</u> needed to make it drinkable.

Drinking Water Needs to Be Good Quality

1) Water's essential for life, but it must be free of <u>poisonous salts</u> (e.g. phosphates and nitrates) and harmful <u>microbes</u>. Microbes in water can cause <u>diseases</u> such as cholera and dysentery.

2) Most of our drinking water comes from <u>reservoirs</u>. Water flows into reservoirs from <u>rivers</u> and <u>groundwater</u> — water companies choose to build reservoirs where there's a good supply of <u>clean water</u>. Government agencies keep a close eye on <u>pollution</u> in reservoirs, rivers and groundwater.

<u>Water from reservoirs goes to the water treatment works for treatment</u>:

1) The water passes though a <u>mesh screen</u> to remove big bits like twigs.

2) Chemicals are added to make solids and microbes <u>stick together</u> and fall to the bottom.

3) The water is <u>filtered</u> through gravel beds to remove all the solids.

4) Water is <u>chlorinated</u> to kill off any harmful <u>microbes</u> left.

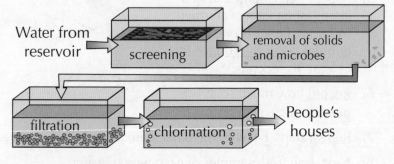

Some people <u>still aren't satisfied</u>. They buy filters that contain <u>carbon</u> or <u>silver</u> to remove substances from their tap water. Carbon in the filters removes <u>chlorine taste</u> and silver is supposed to kill bugs. Some people in hard water areas buy <u>water softeners</u> which contain <u>ion exchange resins</u> (see p.208).

<u>Totally pure water</u> with <u>nothing</u> dissolved in it can be produced by <u>distillation</u> — boiling water to make steam and condensing the steam. This process is too <u>expensive</u> to produce tap water — bags of energy would be needed to boil all the water we use. Distilled water is used in <u>chemistry labs</u>.

You'd use pure water to make a solution of (say) KBr, because you wouldn't want any other ions mucking it up.

Adding Fluoride and Chlorine to Water Has Disadvantages

1) <u>Fluoride</u> is added to drinking water in some parts of the country because it helps to <u>reduce tooth decay</u>. <u>Chlorine</u> is added to <u>prevent disease</u> (see above). So far so good. However...

2) Some studies have linked adding chlorine to water with an <u>increase</u> in certain <u>cancers</u>. Chlorine can <u>react</u> with other <u>natural substances</u> in water to produce <u>toxic by-products</u> which some people think could cause cancer.

3) In <u>high doses</u> fluoride can cause <u>cancer</u> and <u>bone problems</u> in humans, so some people believe that fluoride <u>shouldn't be added</u> to drinking water. There is also concern about whether it's right to 'mass medicate' — people can <u>choose</u> whether to use a <u>fluoride toothpaste</u>, but they can't choose whether their tap water has added fluoride.

4) <u>Levels of chemicals</u> added to drinking water need to be carefully <u>monitored</u>. For example, in some areas the water may already contain a lot of fluoride, so adding more could be harmful.

Detergents

Here's a splendid example of <u>better, cleaner living</u> through Chemistry.

*Detergents have a **Hydrophilic** Head and a **Hydrophobic** Tail*

1) <u>Detergents and soaps</u> have a <u>hydrophobic</u> part and a <u>hydrophilic</u> part.

2) <u>Hydrophobic</u> means it <u>doesn't like water</u>. This part of the molecule is normally a <u>long hydrocarbon chain</u> or 'tail'.

3) <u>Hydrophilic</u> means that something <u>loves water</u>. This part of the molecule is normally <u>small and ionic</u> — the 'head'.

4) The hydrophilic end of the molecule forms strong <u>intermolecular forces</u> with <u>water</u> molecules.

5) The hydrophobic part forms strong intermolecular forces with molecules of <u>oil and fat</u>.

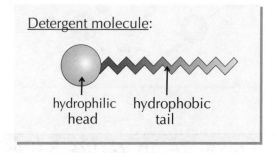

Detergent molecule:

hydrophilic head   hydrophobic tail

6) This means that when detergents come into contact with fat or oil a <u>droplet</u> of oil/fat forms, surrounded by a coating of detergent. This helps <u>lift oily dirt</u> out of fabric — see the diagram.

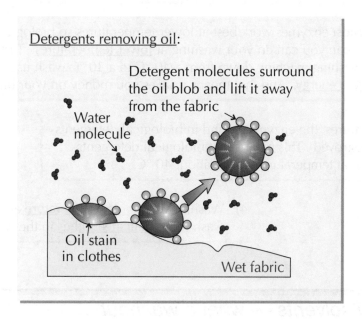

Detergents removing oil:

Detergent molecules surround the oil blob and lift it away from the fabric

Water molecule

Oil stain in clothes

Wet fabric

Detergents and Dry-Cleaning

Water isn't always the best thing for washing — sometimes other solvents are better for removing stains.

Dry Cleaning Uses Solvents to Remove Stains

1) Dry cleaning can be <u>any</u> cleaning process that <u>doesn't use water</u> — other <u>solvents</u> are used instead.

2) These solvents are much <u>better than detergents</u> at removing oil and grease and will clean stains that won't dissolve in water.

3) This is because the solvent can completely <u>dissolve</u> the oil and grease, removing the stains. Here's how it works:

- There are <u>weak intermolecular forces</u> between the solvent molecules. There are also weak intermolecular forces between the molecules of grease.

- When the solvent is applied to the clothes, intermolecular forces are formed <u>between</u> the <u>solvent</u> and <u>grease</u> molecules, so the grease molecules are <u>surrounded</u> by molecules of solvent.

- When the solvent is removed, the <u>grease is removed</u> with it and the clothes are left squeaky clean.

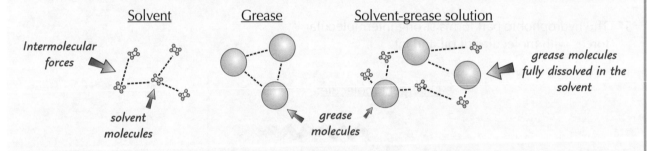

Washing Clothes at a Lower Temperature Saves Energy

1) <u>Biological detergents</u> contain enzymes which are biological <u>catalysts</u>. They help <u>break down</u> some large insoluble molecules into smaller soluble molecules which can be easily removed.

2) Most enzymes work best at lower temperatures so biological detergents mean you can do your washing at <u>lower temperatures</u>. Turning your washing machine down from a 40 °C to a 30 °C wash uses about 40% <u>less energy</u>. And that means it saves you money on your bills... Hooray.

3) At higher temperatures, the enzymes found in biological detergents are <u>denatured</u> (destroyed). This means that biological detergents don't work as well at temperatures above about 40 °C.

4) Washing at a <u>cooler temperature</u> also means you can wash more <u>delicate clothes</u> in the washing machine.

Detergents and solvents — what a washout...

There's lots of clever stuff on this page — so no excuses for not knowing how <u>detergents</u> and <u>dry cleaning solvents</u> work. Washing your clothes at a lower temperature can <u>save energy</u> and <u>money</u>.

Warm-Up and Exam Questions

Just like day follows night, exam questions follow warm-up questions.

Warm-Up Questions

1) What is hard water?
2) What is scale?
3) Name the two types of water hardness.
4) How can totally pure water be produced?
5) What are contained in biological detergents that are not found in other detergents?

Exam Questions

1   Hard water in many areas is caused by dissolved Ca^{2+} ions.

(a)   Give two disadvantages of living in a hard water area.

(2 marks)

(b)   Give two advantages of living in a hard water area.

(2 marks)

(c)   Hard water can be caused by dissolved calcium hydrogencarbonate.
Explain how this type of hardness can be removed by boiling.

(2 marks)

(d)   Give two other methods that can be used to remove hardness.

(2 marks)

2   The diagram below shows how water in some parts of the UK is treated before it reaches people's homes.

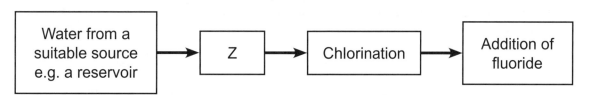

(a)   Describe stage **Z** in the process.

(2 marks)

(b)   *In this question you will be assessed on the quality of your English, the organisation of your ideas and your use of appropriate specialist vocabulary.*

Discuss the advantages and disadvantages of the treatment of water in this way.

(6 marks)

3   Detergent molecules are able to remove dirt, oil and grease from fabrics during washing.

(a)   Explain how the chemical structure of the molecules helps them to remove oil.

(3 marks)

(b)   Some stains have to be removed using dry cleaning. Explain what dry cleaning involves and how it removes difficult stains.

(2 marks)

Fertilisers

There's a lot more to <u>using fertilisers</u> than making your garden look nice and pretty...

Fertilisers Provide Plants with the Essential Elements for Growth

1) The three main <u>essential</u> elements in fertilisers are <u>nitrogen</u>, <u>phosphorus</u> and <u>potassium</u>.

2) If plants don't get enough of these elements, their <u>growth</u> and <u>life processes</u> are affected.

3) These elements may be <u>missing</u> from the soil if they've been <u>used up</u> by a <u>previous crop</u>.

4) Fertilisers <u>replace</u> these missing elements or provide <u>more</u> of them.

5) This helps to increase the <u>crop yield</u>, as the crops can grow <u>faster</u> and <u>bigger</u>.

6) For example, fertilisers add more <u>nitrogen</u> to <u>plant proteins</u>, which makes the plants <u>grow faster</u>.

7) The fertiliser must first <u>dissolve in water</u> before it can be taken in by the crop <u>roots</u>.

Ammonia Can be Neutralised with Acids to Produce Fertilisers

Ammonia is a <u>base</u> and can be <u>neutralised</u> by acids to make <u>ammonium salts</u>. Ammonia is really important to world food production, because it's a <u>key ingredient</u> of many <u>fertilisers</u>.

1) If you neutralise <u>nitric acid</u> with ammonia you get <u>ammonium nitrate</u>. It's an especially good fertiliser because it has <u>nitrogen</u> from <u>two sources</u>, the ammonia and the nitric acid — kind of a <u>double dose</u>.

2) <u>Ammonium sulfate</u> can also be used as a fertiliser. You make it by neutralising <u>sulfuric acid</u> with ammonia.

3) <u>Ammonium phosphate</u> is a fertiliser made by neutralising <u>phosphoric acid</u> with <u>ammonia</u>.

4) <u>Potassium nitrate</u> is also a fertiliser — it can be made by neutralising <u>nitric acid</u> with <u>potassium hydroxide</u>.

Fertilisers

Fertilisers are Useful — But Can Cause Big Problems

1) The population of the world is rising rapidly.

2) Fertilisers increase crop yield, so the more fertiliser we make, the more crops we can grow, and the more people we can feed.

3) But if we use too many fertilisers we risk polluting our water supplies and causing eutrophication.

Fertilisers Damage Lakes and Rivers — Eutrophication

1) When fertiliser is put on fields some of it inevitably runs off and finds its way into rivers and streams.

2) The level of nitrates and phosphates in the river water increases.

3) Algae living in the river water use the nutrients to multiply rapidly, creating an algal bloom (a carpet of algae near the surface of the river). This blocks off the light to the river plants below. The plants cannot photosynthesise, so they have no food and they die.

4) Aerobic bacteria feed on the dead plants and start to multiply. As the bacteria multiply they use up all the oxygen in the water. As a result pretty much everything in the river dies (including fish and insects).

Aerobic just means that they need oxygen to live.

5) This process is called <u>EUTROPHICATION</u>, which basically means 'too much of a good thing'.

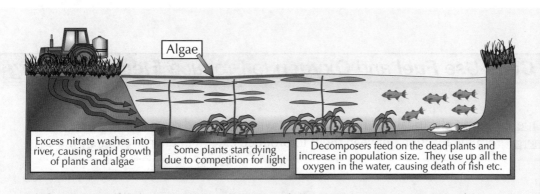

As the picture shows, too many nitrates in the water cause a sequence of 'mega-growth', 'mega-death' and 'mega-decay' involving most of the plant and animal life in the water.

Farmers have to be careful where they're spreading fertilisers...

Unfortunately, no matter how good something is, there's nearly always a downside. If you've ever seen a lake or river covered in a layer of green algae then it's likely that it's caused by eutrophication.

Getting Energy From Hydrogen

Fuel cells are great — they use hydrogen and oxygen to make electricity.

Hydrogen and *Oxygen* Give Out *Energy* When They *React*

1) <u>Hydrogen and oxygen react</u> to produce <u>water</u>.

2) The reaction between hydrogen and oxygen
   is <u>exothermic</u> — it <u>releases energy</u>.

3) You can show this on an <u>energy level diagram</u>.

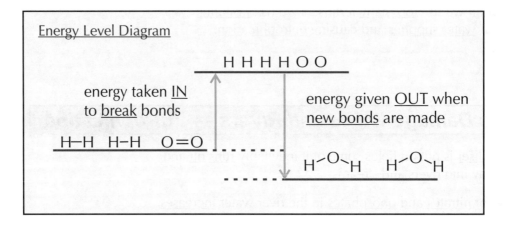

4) The higher a line is on the diagram the <u>more energy</u>
   the substances have. For example the H_2 and O_2
   molecules have a higher energy than the H_2O molecules.

5) When the new bonds are formed the <u>excess energy</u> is given out in the form of heat.

Fuel Cells Use *Fuel* and *Oxygen* to Produce *Electrical Energy*

1) A <u>fuel cell</u> is an electrical cell that's supplied with a <u>fuel</u> and <u>oxygen</u> and uses
   <u>energy</u> from the reaction between them to produce electrical energy <u>efficiently</u>.

2) There are a few <u>different types</u> of fuel cells, using different fuels and different electrolytes.
   One of the types is the <u>hydrogen-oxygen fuel cell</u>.

3) This fuel cell combines hydrogen and oxygen to produce heat <u>energy</u> and nice
   clean <u>water</u>. That means there are no nasty pollutants to worry about.

Getting Energy From Hydrogen

Hydrogen-Oxygen Fuel Cells *Involve a* Redox Reaction

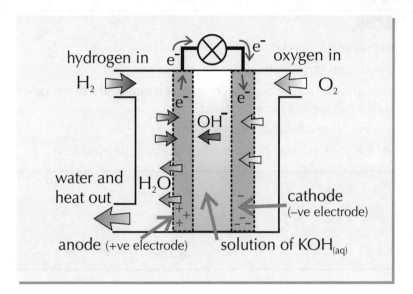

hydrogen in e^- $\otimes$ e^- oxygen in

H_2 O_2

OH^-

water and heat out H_2O

anode (+ve electrode)   cathode (–ve electrode)

solution of $KOH_{(aq)}$

For more about redox reactions see page 154.

1) The electrolyte is often a solution of <u>potassium hydroxide</u>, and the electrodes are often porous carbon with a catalyst.

2) <u>Hydrogen</u> goes into the <u>anode compartment</u> and <u>oxygen</u> goes into the <u>cathode compartment</u>.

3) At the negative <u>cathode</u>, <u>oxygen gains electrons</u> (from the cathode) and reacts with <u>water</u> (from the electrolyte) to make OH⁻ ions.

$$O_2 + 4e^- + 2H_2O \rightarrow 4OH^-$$

4) The oxygen gas is <u>gaining electrons</u> — this is <u>reduction</u>.

5) <u>OH⁻ ions</u> in the electrolyte move to the positive <u>anode</u>.

6) At the positive <u>anode</u>, hydrogen combines with the hydroxide ions to produce <u>water</u> and <u>electrons</u>.

$$2H_2 + 4OH^- \rightarrow 4H_2O + 4e^-$$

7) The hydrogen gas <u>loses electrons</u> — this is <u>oxidation</u>.

8) The electrons <u>flow</u> through an external circuit from the <u>anode</u> to the <u>cathode</u> — this is the <u>current</u>.

9) The overall reaction is <u>hydrogen plus oxygen</u>, which gives <u>water</u>.

$$2H_2 + O_2 \rightarrow 2H_2O \qquad \text{hydrogen + oxygen} \rightarrow \text{water}$$

There's <u>reduction</u> at the cathode and <u>oxidation</u> at the anode, so the whole thing is a <u>REDOX</u> reaction.

Fuel cells don't produce nasty carbon emissions or poisonous gases

In hydrogen-oxygen fuel cells, different things happen at each electrode — oxidation at the anode and reduction at the cathode. The overall reaction produces electrical energy without a load of pollution.

Warm-Up and Exam Questions

You know the idea by now — do the warm-up questions first, then get stuck into the exam questions...

Warm-Up Questions

1) a) How is ammonium phosphate made?
   b) Give one use of ammonium phosphate.
2) Excess nutrients in river water can cause algal bloom, which can result in the death of many river organisms. What name is given to this process?
3) Is the reaction inside a fuel cell exothermic or endothermic?
4) In the hydrogen-oxygen fuel cell, what solution is often used as the electrolyte?

Exam Questions

1   Hydrogen-oxygen fuel cells involve redox reactions.
   (a) Describe what a fuel cell is.
   (1 mark)

   (b) Explain what a redox reaction is.
   (1 mark)

   (c) (i)   One of the half-equations for the hydrogen-oxygen fuel cell is:
   $$O_2 + 4e^- + 2H_2O \rightarrow 4OH^-$$
   Is this an oxidation or a reduction? Explain your choice.
   (1 mark)

   (ii) What is the overall chemical equation for the hydrogen-oxygen fuel cell?
   (2 marks)

   (d) Give one advantage of using a hydrogen-oxygen fuel cell to power a car instead of a traditional petrol engine.
   (1 mark)

2   Farmers often use fertilisers.
   (a) Explain why fertilisers are used by farmers.
   (3 marks)

   (b) Name two elements that fertilisers usually contain.
   (2 marks)

   (c) Ammonium nitrate is a fertiliser made by neutralising nitric acid (HNO_3) and ammonia (NH_3). Suggest why ammonium nitrate is a particularly good fertiliser.
   (1 mark)

   (d) Name the fertiliser that can be made by neutralising sulfuric acid with ammonia.
   (1 mark)

   (e) Fertilisers can cause problems such as eutrophication. Eutrophication happens when the level of nitrates and phosphates in water sources increases.
   Outline the process of eutrophication.
   (6 marks)

Revision Summary for Section 11

Bit of a mixed bag — one minute you're pondering water hardness, the next you're worrying about the effects of eutrophication on all of those poor fish. That's one of the many joys of Chemistry.
Test yourself on these little beauties. And if you don't know any, flick back and learn the stuff again.

1) What are the main ions that cause water hardness?

2) Give two possible health benefits of drinking hard water.

3) Explain how adding washing soda to hard water removes the hardness.

4) Describe how you could use titration to compare the hardness of two different water samples.

5) During water treatment, how are microbes killed so that the water is safe to drink?

6) Explain why tap water isn't purified by distillation.

7) Why is fluoride added to some drinking water?

8) What does hydrophilic mean? What does hydrophobic mean?

9) What is a biological detergent?

10) Give one advantage of using a biological detergent to wash clothes rather than a standard detergent.

11) How does nitrogen increase the growth of plants?

12) Name two fertilisers which are manufactured from ammonia.

13) Describe what can happen if too much fertiliser is put onto fields.

14) Sketch an energy level diagram for the reaction between hydrogen and oxygen.

15) In a hydrogen-oxygen fuel cell, is the hydrogen oxidised or reduced?

16) Write down the half-equations for the oxidation and reduction reactions that take place inside a hydrogen-oxygen fuel cell.

The Chemical Industry

Absolutely loads of some types of chemicals are used — such as <u>fertilisers</u>. Well they don't just grow on trees. No, they have to be <u>made</u>. And made they are — on a <u>massive scale</u>.

Some Chemicals are **Produced** on a **Large Scale**...

There are certain chemicals that industries need <u>thousands and thousands of tonnes</u> of every year — <u>ammonia</u>, <u>sulfuric acid</u>, <u>sodium hydroxide</u> and <u>phosphoric acid</u> are four examples.
Chemicals that are produced on a <u>large scale</u> are called <u>bulk chemicals</u>.

...And Some are **Produced** on a **Small Scale**

1) Some chemicals aren't needed in such large amounts — but that doesn't mean they're any less important.

2) Chemicals produced on a <u>smaller scale</u> are called <u>fine chemicals</u>.

3) Some examples are <u>drugs</u>, <u>food additives</u> and <u>fragrances</u>.

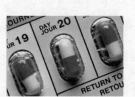

New **Chemical Products** Need Lots of **Research**

Before new chemical products are made, a huge amount of <u>research and development</u> work goes on. This can take <u>years</u>, and be <u>really expensive</u>, but it's worth it in the end if the company makes lots of <u>money</u> out of the new product.

For example, to make a new production process run efficiently a new <u>catalyst</u> might have to be found. This is likely to involve:

1) <u>Testing</u> potential catalysts using a process of <u>trial and error</u>.

2) Making <u>computer models</u> of the reaction to try to work out which substance might work as a catalyst.

3) <u>Designing or refining the manufacture</u> of the catalyst to make sure that the new product can be mass-produced <u>safely</u>, <u>efficiently</u>, and <u>cost effectively</u>.

4) Investigating the <u>risks to the environment</u> of using the new catalyst and trying to <u>minimise</u> them.

5) <u>Monitoring</u> the <u>quality</u> of the product to make sure that it is not affected by the catalyst.

These jobs, and lots of other types of work, are done by people in the <u>chemical industry</u>.

Fine chemicals — by appointment to Her Majesty, The Queen...

Producing <u>bulk</u> chemicals is like painting a house — it's huge, so you slap that paint on with a <u>great big</u> brush. Producing <u>fine</u> chemicals is like painting a picture — much <u>smaller</u> and more <u>fiddly</u>.

The Chemical Industry and Regulations

Chemicals can be <u>dangerous</u> not only to us, but also to the world around us. Luckily, the <u>government</u> have created a set of <u>regulations</u> to make sure everything is kept under control.

Government Regulations Protect People and the Environment

Governments place <u>strict controls</u> on everything to do with <u>chemical processes</u>. This is done to <u>protect</u> <u>workers</u>, the <u>general public</u> and the <u>environment</u>.

For example, there are regulations about...

Using chemicals

E.g. <u>sulfuric acid</u> is sprayed on potato fields to <u>destroy</u> the leaves and stalks of the potato plants and make harvesting easier.

Government <u>regulations restrict</u> how much acid can be used and require <u>signs</u> to be displayed to <u>warn</u> the public.

Storage

Many <u>dangerous chemicals</u> have to be stored in locked storerooms. <u>Poisonous</u> chemicals must be stored in either <u>sealed containers</u> or well-ventilated store cupboards.

Transport

E.g. lorries transporting chemicals must display <u>hazard symbols</u> and <u>identification numbers</u> to help the emergency services deal <u>safely</u> with any accidents and spills.

The regulations are there to keep us safe

You should feel reassured that there are regulations in place to stop people splashing round dangerous chemicals whenever and however they feel like it. I think we'll all sleep better for knowing that...

Life Cycle Assessments

If a company wants to manufacture a new product, they carry out a <u>life cycle assessment (LCA)</u>. This looks at every <u>stage</u> of the product's life to assess the <u>impact</u> it would have on the environment.

Life Cycle Assessments Show Total Environmental Costs

A <u>life cycle assessment (LCA)</u> looks at each <u>stage</u> of the <u>life</u> of a product — from making the <u>material</u> from natural raw materials (see page 225), making the <u>product</u> from the material, <u>using</u> the product and <u>disposing</u> of the product. It works out the potential <u>environmental impact</u> of each stage.

CHOICE OF MATERIAL

1) Most chemical manufacture needs <u>water</u>.
2) <u>Metals</u> have to be <u>mined</u> and <u>extracted</u> from their ores. These processes need a lot of <u>energy</u> and cause a lot of <u>pollution</u>.
3) <u>Raw materials</u> for chemical manufacture often come from <u>crude oil</u>. Crude oil is a <u>non-renewable resource</u>, and supplies are <u>decreasing</u> as oil is a <u>finite resource</u>. Also, obtaining crude oil from the ground and refining it into useful raw materials requires a lot of <u>energy</u> and generates <u>pollution</u>.

MANUFACTURE

1) <u>Manufacturing</u> products uses a lot of <u>energy</u> and other resources.
2) It can also cause a lot of <u>pollution</u>, e.g. <u>harmful fumes</u> such as CO or HCl.
3) You also need to think about any <u>waste</u> products and how to <u>dispose</u> of them.
4) Some waste can be <u>recycled</u> and turned into other <u>useful chemicals</u>, reducing the amount that ends up polluting the environment.

Some products can be recycled — the materials can be processed and used again in new products.

PRODUCT DISPOSAL

1) Products are often <u>disposed</u> of in a <u>landfill</u> site at the end of their life.
2) This takes up space and <u>pollutes</u> land and water, e.g. when paint washes off a product and gets into rivers.
3) Products might be <u>incinerated</u> (burnt), which causes air pollution.

USING THE PRODUCT

<u>Using</u> the product can also damage the environment. For example:
1) <u>Paint</u> gives off <u>toxic fumes</u>.
2) <u>Burning fuels</u> releases <u>greenhouse gases</u> and other <u>harmful substances</u>.
3) <u>Fertilisers</u> can <u>leach</u> into streams and rivers and cause damage to <u>ecosystems</u>.

Need exercise? Go life-cycling then...

Life cycle assessments are really good for <u>evaluating</u> the different materials you could use for a particular product. Once you know what needs to be considered at <u>each stage</u> you can evaluate the materials <u>one by one</u>. And you can also evaluate the <u>product</u> as a whole.

Synthesising Compounds

As I'm sure you know, the chemical industry is really important. Without it we'd be without loads of everyday chemicals. When it comes to making these chemicals, there are quite a few stages to the process. The seven stages are covered over the next two pages.

There are **Seven Stages** Involved in **Chemical Synthesis**

1) Choosing the Reaction

Chemists need to choose the reaction (or series of reactions) to make the product.

For example:
- neutralisation (see p. 145) — an acid and an alkali react to produce a salt.
- thermal decomposition — heat is used to break up a compound into simpler substances.
- precipitation — an insoluble solid is formed when two solutions are mixed.

2) Risk Assessment

This is an assessment of anything in the process that could cause injury.

It involves:
- identifying hazards.
- assessing who might be harmed.
- deciding what action can be taken to reduce the risk.

3) Calculating the Quantities of Reactants

1) This includes a lot of maths and a balanced symbol equation.

2) Using the equation chemists can calculate how much of each reactant is needed to produce a certain amount of product.

3) This is particularly important in industry because you need to know how much of each raw material is needed so there's no waste — waste costs money.

Synthesising Compounds

4) Choosing the Apparatus and Conditions

The reaction needs to be carried out using suitable <u>apparatus</u> and in the right <u>conditions</u>.

For example:

The apparatus needs to be the <u>correct size</u> (for the <u>amount</u> of product and reactants) and <u>strength</u> (for the <u>type</u> of reaction being carried out, e.g. if it is explosive or gives out a lot of heat).

Chemists need to decide what <u>temperature</u> the reaction should be carried out at, what <u>concentrations</u> of reactants should be used, and whether or not to use a <u>catalyst</u> (see p. 167-169).

5) Isolating the Product

After the reaction is finished the products may need to be separated from the <u>reaction mixture</u>.

This could involve:

- <u>evaporation</u> (if the product is dissolved in the reaction mixture).
- <u>filtration</u> (if the product is an <u>insoluble solid</u>).
- <u>drying</u> (to remove any <u>water</u>).

6) Purification

1) Isolating the product and <u>purification</u> go together like peas and carrots. As you're isolating the product you're also helping to purify it.

2) <u>Crystallisation</u> can be useful in the purification process.

7) Measuring Yield and Purity

The yield tells you about the <u>overall success</u> of the process. It compares what you think you should get with what you get in practice (see p. 140). The purity of the chemical also needs to be measured (p. 196).

It's just like Snow White — but with chemical synthesis steps...

It's important that none of these stages are <u>missed out</u>. It'd be pointless if you weren't able to <u>separate</u> the product from the <u>reaction mixture</u> and even worse if the process caused <u>injury</u> or <u>death</u>. Be safe.

Producing Chemicals

Producing chemicals is a complicated business. Luckily most processes involve the same stages.

There Are **Several Stages** Involved in **Producing Chemicals**

The process of producing a useful chemical from the raw materials can be split into <u>five</u> stages:

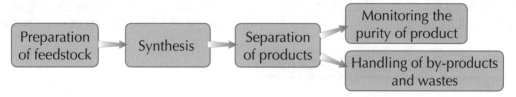

Preparation of feedstock → Synthesis → Separation of products → Monitoring the purity of product / Handling of by-products and wastes

1) **Raw Materials** are Converted Into **Feedstocks**

1) <u>Raw materials</u> are the <u>naturally</u> occurring substances which are needed, e.g. crude oil, natural gas.
2) <u>Feedstocks</u> are the actual <u>reactants</u> needed for the process, e.g. hydrogen, ethanol.
3) The raw materials usually have to be <u>purified</u> or changed in some way to make the feedstock.

2) **Synthesis**

The <u>feedstocks</u> (reactants) are converted by the magic of chemistry into <u>products</u>. The <u>conditions</u> have to be carefully controlled to make sure the <u>reaction</u> happens, and at a sensible <u>rate</u>.

3) The **Products** are **Separated**

1) Chemical reactions usually produce the <u>substance you want</u> and some <u>other chemicals</u> known as <u>by-products</u>. The by-products might be useful, or they might be <u>waste</u>.
2) You might also have some <u>left-over reactants</u>.
3) Everything has to be <u>separated out</u> so it can be dealt with in different ways.

4) The **Purity** of the Product is **Monitored**

1) Even after the best efforts are made to separate the product from everything else, it sometimes still has other things <u>mixed in</u> with it — it's not completely <u>pure</u>.
2) The <u>purity</u> of the product has to be <u>monitored</u> to make sure it's between certain levels.
3) Different industries need <u>different levels</u> of purity depending on what the product is <u>used</u> for. If a <u>slightly impure</u> product will do the <u>job</u> it's meant for, there's no point <u>wasting money</u> on purification.

5) **By-products** and **Waste** are Dealt With

1) Where possible, <u>by-products</u> are <u>sold</u> or used in <u>another reaction</u>.
2) If the reaction is <u>exothermic</u>, there may be <u>waste heat</u>. <u>Heat exchangers</u> can use excess heat to produce steam or hot water for <u>other reactions</u> — saving energy and money.
3) Waste products have to be <u>carefully disposed of</u> so they don't harm people or the environment — there are <u>legal requirements</u> about this.

Producing Chemicals

It'd be great if all industrial reactions were <u>sustainable</u> — humans could go <u>on</u> making whatever they wanted <u>forever</u> and ever. Life <u>isn't</u> like that though — so it's important to think about sustainability.

*There Are **Eight** Key **Questions** About **Sustainability***

<u>Sustainable processes</u> are ones that <u>meet people's needs today without affecting the ability of future generations to meet their own needs</u>. Lots of factors affect whether a chemical process is sustainable.

1) Will The Raw Materials Run Out?

1) It's great if your feedstock is <u>renewable</u> — you can keep on using as much as you like.

2) The trouble is, if it's not renewable it's going to <u>run out</u>. And this could mean <u>big problems</u> for future generations.

2) How Good is The Atom Economy?

1) The <u>atom economy</u> of a reaction tells you how much of the <u>mass</u> of the reactants ends up as useful products.

2) Pretty obviously, if you're making <u>lots of waste</u>, that's a <u>problem</u> — it all has to go somewhere.

3) Reactions with low atom economy <u>use up resources</u> very quickly too.

3) What Do I Do With The Waste Product?

1) Waste products can be expensive to <u>remove</u> and dispose of <u>responsibly</u>. They're likely to <u>take up space</u> and cause <u>pollution</u>.

2) One way around the problem is to find a <u>use</u> for the waste products rather than just <u>throwing them away</u>.

3) Alternatively, there's often <u>more than one way</u> to make the product you want, so you try to choose a reaction that gives <u>useful by-products</u>.

4) What Are The Energy Costs?

1) If a reaction needs a lot of <u>energy</u> it'll be very <u>expensive</u>.

2) And providing energy often involves <u>burning fossil fuels</u> — which of course is no good for the <u>environment</u>.

3) But if a process <u>gives out</u> energy there might be a way to <u>use</u> that energy for something else — <u>saving money</u> and the <u>environment</u>.

Producing Chemicals

5) Will it Damage The Environment?

1) Clearly if the reaction produces <u>harmful chemicals</u> it's <u>not</u> going to do any good for the <u>environment</u>.

2) But you need to consider where the <u>raw materials</u> come from too (mining, for example, can make a right mess of the countryside, see p. 42), and also whether the reactants or products need <u>transporting</u>.

6) What Are The Health and Safety Risks?

There's no doubt about it — chemistry can be <u>dangerous</u>.

1) There are <u>laws</u> in place that companies must follow to make sure their workers and the public are <u>not</u> put in harm's way.

2) Companies also have to <u>test</u> their products to make sure they're <u>safe to use</u>.

7) Are There Any Benefits or Risks To Society?

A factory creates <u>jobs</u> for the local community and brings <u>money</u> into the area. But it may be <u>unsightly</u> and potentially <u>hazardous</u>.

8) Is It Profitable?

This is the big question for most companies — businesses are out to make money after all. If the <u>costs</u> of a process are <u>higher</u> than the <u>income</u> from it, then it won't be <u>profitable</u>.

Enough with the questions — I confess...

If you want to make a product safely and sustainably than you have to answer these eight questions. If this is still all a bit confusing there are some <u>examples</u> of how to go about it on p. 231 and 235.

Warm-Up and Exam Questions

So many questions. Questions to ask when producing chemicals, questions about sustainability, warm-up questions and exam questions. Is there ever an end to the questions? Yes.

Warm-Up Questions

1) Give one example of a chemical produced on a large scale.

2) Give one example of a chemical produced on a small scale.

3) What does an LCA show?

4) What is a raw material?

Exam Questions

1   A business in the chemical industry is considering which of two processes (A or B) it will use to manufacture a product.

   (a)   Process A has a lower atom economy than process B.

      (i)   Explain what is meant by the term atom economy.

(1 mark)

      (ii)   Give two disadvantages of using a chemical process with a low atom economy.

(2 marks)

   (b)   The product is a bulk chemical. Explain what a bulk chemical is.

(1 mark)

   (c)   The feedstocks used for process B are renewable.
Explain what is meant by the term renewable feedstock.

(2 marks)

   (d)   Governments place strict controls on the processes involved in producing chemicals.
Using examples, outline two ways that governments regulate chemical production to protect people.

(4 marks)

   (e)   The business carries out a risk assessment for process B.
Describe what a risk assessment is and how one is carried out.

(4 marks)

2   The sustainability of a product is an important consideration for manufacturers.

   (a)   Describe what a sustainable process is.

(1 mark)

   (b)   *In this question you will be assessed on the quality of your English, the organisation of your ideas and your use of appropriate specialist vocabulary.*

Chemical processes produce waste, both during the manufacture of products and at the end of a product's useful life.

Discuss how the disposal of waste materials can affect a product's sustainability.

(6 marks)

The Haber Process

This is an important industrial process. It produces ammonia (NH_3), which is used to make fertilisers.

Nitrogen and Hydrogen are Needed to Make Ammonia

$$N_2(g) + 3H_2(g) \rightleftharpoons 2NH_3(g) \quad (+ \text{ heat})$$

1) The nitrogen is obtained easily from the air, which is 78% nitrogen (and 21% oxygen). *These gases are first purified.*

2) The hydrogen comes from natural gas or from other sources like crude oil.

3) Some of the nitrogen and hydrogen reacts to form ammonia. Because the reaction is reversible — it occurs in both directions — ammonia breaks down again into nitrogen and hydrogen. The reaction reaches an equilibrium.

> **Industrial conditions**: pressure = 200 atmospheres; temperature = 450 °C; catalyst: iron.

The Reaction is Reversible, So There's a Compromise to be Made:

1) Higher pressures favour the forward reaction (since there are four molecules of gas on the left-hand side, for every two molecules on the right — see the equation above).

2) So the pressure is set as high as possible to give the best % yield, without making the plant too expensive to build (it'd be too expensive to build a plant that'd stand pressures of over 1000 atmospheres, for example). Hence the 200 atmospheres operating pressure.

3) The forward reaction is exothermic, which means that increasing the temperature will actually move the equilibrium the wrong way — away from ammonia and towards N_2 and H_2. So the yield of ammonia would be greater at lower temperatures.

See pages 178-180 for more on equilibrium.

4) The trouble is, lower temperatures mean a lower rate of reaction. So what they do is increase the temperature anyway, to get a much faster rate of reaction.

5) The 450 °C is a compromise between maximum yield and speed of reaction. It's better to wait just 20 seconds for a 10% yield than to have to wait 60 seconds for a 20% yield.

6) The ammonia is formed as a gas but as it cools in the condenser it liquefies and is removed.

7) The unused hydrogen (H_2) and nitrogen (N_2) are recycled so nothing is wasted.

H_2 and N_2 mixed in 3:1 ratio

Reaction vessel

Trays of iron catalyst

450 °C
200 atm

Unused H_2 and N_2 is recycled

Condenser

Liquid Ammonia

The Iron Catalyst Speeds Up the Reaction and Keeps Costs Down

1) The iron catalyst makes the reaction go faster, which gets it to the equilibrium proportions more quickly. But remember, the catalyst doesn't affect the position of equilibrium (i.e. the % yield).

2) Without the catalyst the temperature would have to be raised even further to get a quick enough reaction, and that would reduce the % yield even further. So the catalyst is very important.

Nitrogen Fixation

Nitrogen fixation is about taking nitrogen and turning it into chemicals we can use.
Just like in the Haber process...

Nitrogen Fixation Turns N₂ from the Air into Ammonia

1) Nitrogen fixation is the process of turning N_2 from the air into useful
nitrogen compounds like ammonia.

2) The Haber process is a non-biological way of fixing nitrogen (see previous page).

3) Most of the ammonia produced by the Haber process is used to make fertilisers.

4) Fertilisers play a vital part in world food production as they increase
crop yield so help to feed more people.

5) When used in large amounts though, fertilisers can pollute water supplies
and cause eutrophication (see page 215).

6) Ammonia is also very important in industry where it is used to manufacture
plastics, explosives and pharmaceuticals.

The Efficiency of Nitrogen Fixation can be Improved by Catalysts

1) In the Haber process very high temperatures and pressures
have to be used to turn nitrogen and hydrogen into ammonia.

2) Using an iron catalyst makes the rate of reaction
much faster — so the ammonia is produced faster.

3) Without the catalyst the temperature would have to be raised even further to get a quick enough
reaction, and that would reduce the % yield even further. So the catalyst is very important.

4) Some living organisms such as nitrogen-fixing bacteria can fix nitrogen at room
temperature and pressure. They do this using biological catalysts called enzymes.

5) Chemists would like to be able to make catalysts that mimic these enzymes, so that
processes like the Haber process can be carried out at room temperature and pressure.

6) As it's expensive and time consuming to work at high temperatures
and pressures, this would mean that processes involving nitrogen
fixation would become much cheaper and more efficient.

Nitrogen Fixation

Is **Nitrogen Fixation** A **Sustainable** Process?

Here's a run down of the sustainability of the Haber process.

1) Will the raw materials run out?

Hydrogen comes from <u>fossil fuels</u>. They're <u>non-renewable</u> and will run out. Nitrogen comes from the air so it's very unlikely it will run out.

2) How good is the atom economy?

<u>All</u> the H_2 and N_2 makes ammonia so the atom economy is <u>excellent</u>.

3) What do I do with my waste products?

There are no waste products as the chemicals are all <u>recycled</u>.

4) What are the energy costs?

Lots of <u>energy</u> is needed to keep the reaction at 450 °C and 200 atm.

5) Will it damage the environment?

<u>Fertilisers</u> made from NH_3 can cause <u>eutrophication</u> and <u>water pollution</u>.

6) What are the health and safety risks?

Working at <u>high temperatures</u> and <u>pressures</u> can be very dangerous.

7) Are there any benefits or risks to society?

Making ammonia can help <u>world food production</u>.

8) Is it profitable?

Yes, making ammonia is <u>big business</u>.

Nitrogen-fixing bacteria — they've got it sorted...

If I could invent a process half as efficient as those <u>nitrogen-fixing bacteria</u> I'd have more money than I could shake a stick at. Shame — I'll have to come up with another way to make my millions.

The Contact Process

And here's another example where getting the <u>conditions</u> right makes you <u>more product</u>.

The **Contact Process** is Used to Make **Sulfuric Acid**

(1) The first stage is to make <u>sulfur dioxide</u> (SO_2) — usually by burning <u>sulfur</u> in <u>air</u>.

> sulfur + oxygen → sulfur dioxide
>
> $S_{(s)} + O_{2(g)} \rightarrow SO_{2(g)}$

(2) The sulfur dioxide is then <u>oxidised</u> (with the help of a catalyst) to make <u>sulfur trioxide</u> (SO_3).

> sulfur dioxide + oxygen ⇌ sulfur trioxide
>
> $2SO_{2(g)} + O_{2(g)} \rightleftharpoons 2SO_{3(g)}$

(3) Next, the sulfur trioxide is used to make <u>sulfuric acid</u>.

> sulfur trioxide + water → sulfuric acid
>
> $SO_{3(g)} + H_2O_{(l)} \rightarrow H_2SO_{4(aq)}$

In reality, dissolving SO_3 like this doesn't work — the reaction is dangerous as a lot of heat's produced. (In practice, you dissolve SO_3 in sulfuric acid first.)

The **Conditions** Used to Make SO_3 are **Carefully Chosen**

The reaction in step 2 is <u>reversible</u>. So, the <u>conditions</u> used can be <u>controlled</u> to get a <u>higher yield</u> (more product).

$$2SO_2 + O_2 \rightleftharpoons 2SO_3$$

The forward reaction is exothermic.

Temperature

1) Oxidising sulfur dioxide to form sulfur trioxide is <u>exothermic</u> (it <u>gives out</u> heat).
2) So to get <u>more product</u> you'd think the temperature should be <u>reduced</u> (so the equilibrium will shift to the <u>right</u> to <u>replace the heat</u>).
3) Unfortunately, reducing the temperature <u>slows</u> the reaction right down — not much good.
4) So an <u>optimum</u> temperature of <u>450 °C</u> is used, as a compromise.

Pressure

1) There are <u>two moles</u> of <u>product</u>, compared to <u>three moles</u> of <u>reactants</u>.
2) So to get <u>more product</u>, you'd think the pressure should be <u>increased</u> (so that the equilibrium will shift to the <u>right</u> to <u>reduce the pressure</u>).
3) But increasing the pressure is <u>expensive</u>, and as the equilibrium is already on the right, it's not really <u>necessary</u>.
4) In fact, <u>atmospheric pressure</u> (1 atmosphere) is used.

Catalyst

1) To <u>increase</u> the rate of reaction a <u>vanadium pentoxide catalyst</u> (V_2O_5) is used.
2) This <u>DOESN'T</u> change the <u>position</u> of the equilibrium.

> With a <u>fairly high temperature</u>, a <u>low pressure</u> and a <u>vanadium pentoxide catalyst</u>, the reaction goes <u>pretty quickly</u> and you get a <u>good yield</u> of SO_3 (about 99%).

Making Ethanol

Ethanol is the alcohol people <u>drink</u> — but as you saw on pages 74-75 this is far from its only use. It's also used as a <u>fuel</u>, a <u>solvent</u> and as a <u>feedstock</u> for other processes.

Ethanol can be Made by Fermentation

The ethanol in <u>alcoholic drinks</u> is usually made using fermentation.

1) <u>Fermentation</u> uses <u>yeast</u> to convert <u>sugars</u> into <u>ethanol</u>. Carbon dioxide is also produced.

$$\text{sugar} \xrightarrow{\text{yeast}} \text{ethanol} + \text{carbon dioxide}$$

2) The yeast cells contain <u>zymase</u>, an <u>enzyme</u> that acts as a catalyst in fermentation.

3) Fermentation happens fastest at about <u>30 °C</u>. That's because zymase works best at this temperature. At lower temperatures, the reaction slows down. And if it's <u>too hot</u> the zymase is <u>denatured</u> (destroyed).

4) Zymase also works best at a <u>pH</u> of about <u>4</u> — a strongly acidic or alkaline solution will stop it working.

5) It's important to prevent <u>oxygen</u> getting to the fermentation process. Oxygen converts the <u>ethanol</u> to <u>ethanoic acid</u> (the acid in <u>vinegar</u>), which lowers the pH and can stop the enzyme working.

6) When the <u>concentration</u> of ethanol reaches about 10 to 20%, the fermentation reaction <u>stops</u>, because the yeast gets <u>killed off</u> by the ethanol.

Ethanol Solution can be Concentrated by Distillation

The fermented mixture can be <u>distilled</u> to produce more <u>concentrated</u> ethanol, which can then be used to make products like <u>brandy</u> or <u>whisky</u>.

1) The ethanol solution is put in a flask below a <u>fractionating column</u>, as shown.

2) The solution is <u>heated</u> so that the <u>ethanol</u> boils. The ethanol vapour travels <u>up</u> the column, <u>cooling</u> down as it goes.

3) The temperature is such that anything with a <u>higher</u> boiling point than ethanol (like water) cools to a <u>liquid</u> and flows <u>back</u> into the <u>solution</u> at the bottom.

4) This means that <u>only</u> pure ethanol vapour reaches the <u>top</u> of the column.

5) The ethanol vapour flows through a <u>condenser</u> — where it's cooled to a <u>liquid</u>, which is then collected.

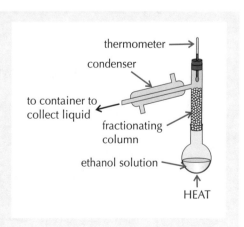

Making Ethanol

Fermentation of sugars isn't the only way of making ethanol. Here are two more to read about.

Ethanol can be Made From Biomass

Scientists have recently developed a way to make <u>ethanol</u> from <u>waste biomass</u>.

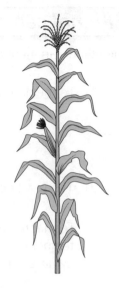

1) <u>Waste biomass</u> is the parts of a plant that would normally be <u>thrown away</u> — e.g. <u>corn stalks</u>, <u>rice husks</u>, <u>wood pulp</u> and <u>straw</u>.

2) Waste biomass <u>cannot</u> be fermented in the <u>normal</u> way because it contains a lot of <u>cellulose</u>.

3) Yeast can easily convert some sugars to ethanol, but it <u>can't</u> convert <u>cellulose</u> to <u>ethanol</u>.

4) <u>E. coli bacteria</u> can be genetically modified so they <u>can</u> convert cellulose in waste biomass into <u>ethanol</u>.

5) The <u>optimum conditions</u> for this process are a temperature of <u>35 °C</u> and a <u>slightly acidic</u> solution, <u>pH 6</u>.

Ethene Can be Reacted with Steam to Produce Ethanol

1) <u>Fermentation</u> is <u>too slow</u> for making ethanol on a <u>large scale</u>.

2) Instead, ethanol is made on an <u>industrial scale</u> using <u>ethane</u>.

3) This method allows <u>high quality</u> ethanol to be produced <u>continuously and quickly</u>.

- <u>Ethane</u> is one of the <u>hydrocarbons</u> found in <u>crude oil</u> and <u>natural gas</u>.

- It is '<u>cracked</u>' (split) to form <u>ethene</u> (C_2H_4) and <u>hydrogen gas</u>.

  ethane → ethene + hydrogen

- <u>Ethene</u> will react with <u>steam</u> (H_2O) to make <u>ethanol</u>.

  ethene + steam → ethanol

- The reaction needs a <u>temperature</u> of 300 °C and a <u>pressure</u> of 70 atmospheres. <u>Phosphoric acid</u> is used as a <u>catalyst</u>.

Sustainability of Making Ethanol

You can use the questions from pages 226-227 to compare the different ways of making ethanol.

Is *Fermentation A Sustainable Process?*

1) <u>Will the raw materials run out?</u> — Sugar beet and yeast <u>grow quickly</u> so <u>won't</u> run out.

2) <u>How good is the atom economy?</u> — The <u>waste CO_2</u> produced means it has a <u>low atom economy</u>. And because the enzyme is <u>killed off</u> by the ethanol produced, the reaction is even <u>less</u> efficient.

3) <u>What do I do with my waste products?</u> — The <u>waste CO_2</u> can be <u>released</u> without any processing.

4) <u>What are the energy costs?</u> — <u>Energy</u> is needed to keep the reaction at its <u>optimum</u> temperature.

5) <u>Will it damage the environment?</u> — <u>Carbon dioxide</u> is a <u>greenhouse gas</u> so adds to <u>global warming</u>.

6) <u>What are the health and safety risks?</u> — The chemicals and processes do <u>not</u> have any specific dangers.

7) <u>Are there any benefits or risks to society?</u> — Making ethanol <u>doesn't</u> impact society (drinking it does).

8) <u>Is it profitable?</u> — This depends on what the ethanol is <u>used</u> for, e.g. <u>drinking</u> or <u>fuel</u>.

Is Producing Ethanol from **Biomass** a **Sustainable** Process?

The <u>sustainability</u> of the <u>biomass</u> method is <u>very similar</u> to the sustainability of the <u>standard</u> <u>fermentation</u> method because they both use <u>similar processes</u>. The <u>advantage</u> of using <u>biomass</u> is that you <u>don't</u> have to <u>grow crops</u> specially for producing ethanol — you can use the <u>waste</u> from other crops.

Is Producing Ethanol from **Ethane** a **Sustainable** Process?

1) <u>Will the raw materials run out?</u> — <u>Crude oil</u> and <u>natural gas</u> are <u>non-renewable</u> and will <u>run out</u>.

2) <u>How good is the atom economy?</u> — <u>Cracking</u> ethane has a <u>fairly high</u> atom economy as the only <u>waste product</u> is hydrogen. Reacting ethene has an <u>even higher atom economy</u> as ethanol is the only product.

3) <u>What do I do with my waste products?</u> — The only waste is the <u>hydrogen gas</u> produced by cracking ethane. It can be <u>reused</u> to make <u>ammonia</u> in the <u>Haber process</u>.

4) <u>What are the energy costs?</u> — <u>Energy</u> is needed to maintain the <u>high temperature</u> and <u>pressure</u> used.

5) <u>Will it damage the environment?</u> — The reactions involved do <u>not</u> produce any <u>waste products</u> that <u>directly</u> harm the environment. But, <u>crude oil</u> can harm the environment, e.g. through oil spills.

6) <u>What are the health and safety risks?</u> — The <u>high temperature and pressure</u> used to produce the ethanol have to be carefully controlled — otherwise it could be <u>very dangerous</u>.

7) <u>Are there any benefits or risks to society?</u> — This method has <u>no</u> specific impact on <u>society</u>.

8) <u>Is it profitable?</u> — <u>Yes</u>, manufacturing ethanol from ethene and steam is <u>continuous</u> and <u>quick</u> and the raw materials are <u>fairly cheap</u> — but it won't stay that way once <u>crude oil</u> starts to run out.

Biomass — an organic church...

It's important to think about the <u>sustainability</u> of chemical processes — it'd be rather selfish to use up all the raw materials, and leave future generations with big piles of our waste to put up with instead.

Warm-Up and Exam Questions

And now onto the best part of the section — the warm up and exam questions. Go for it.

Warm-Up Questions

1) Name the catalyst used in the Haber process.

2) What happens to leftover reactants that are not converted to product in the Haber process?

3) What catalyst is used in the manufacture of sulfuric acid?

4) Name the process that uses a fractionating column to make concentrated ethanol.

Exam Questions

1   Ethanol can be made using fermentation.

   (a) (i)   Complete the equation showing the process of fermentation.

   $$\underline{\hspace{3cm}} \xrightarrow{\text{yeast}} \underline{\hspace{2.5cm}} + \underline{\hspace{2.5cm}}$$

(2 marks)

   (ii) Name the enzyme in yeast that acts as a catalyst in fermentation.

(1 mark)

   (iii) Fermentation works fastest at 30 °C and around pH 4.
         Explain why this is.

(2 marks)

   (iv) Explain why it is important to prevent oxygen from entering the
         fermentation process.

(2 marks)

   (b) Ethanol produced by fermentation is distilled using the apparatus shown.

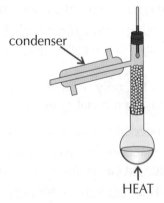

   (i)   Describe the process used to distil ethanol.

(3 marks)

   (ii) Explain why the ethanol produced by fermentation is distilled.

(2 marks)

   (c) Give one use for the ethanol produced by fermentation.

(1 mark)

Exam Questions

2   Ethanol can be produced by reacting ethene with steam.

(a)   Explain how the ethene used in this process is obtained.

(2 marks)

(b)   State the conditions used for this reaction.

(3 marks)

(c)   *In this question you will be assessed on the quality of your English, the organisation of your ideas and your use of appropriate specialist vocabulary.*
Discuss the sustainability of this process.

(6 marks)

3   In industry, sulfuric acid is made from sulfur in a multi-step process.

(a)   Name this process.

(1 mark)

(b)   One step in the process involves a reversible reaction:

$$2SO_2(g) + O_2(g) \rightleftharpoons 2SO_3(g)$$

(i)   The forward reaction is exothermic and so low temperatures would favour a high yield, but the reaction vessel is kept at about 450 °C. Explain why.

(1 mark)

(ii)   What pressure is the reaction vessel kept at? Explain why this is chosen.

(3 marks)

4   Ammonia, NH_3, is made by combining nitrogen and hydrogen at a pressure of 200 atm, a temperature of 450 °C and in the presence of a catalyst.
A flow diagram is shown for the reaction:

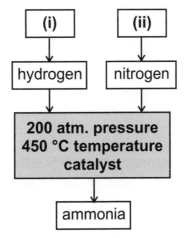

(a)   Write labels for boxes (i) and (ii) to show the sources of nitrogen and hydrogen.

(2 marks)

(b)   Write a balanced equation with state symbols for the reaction between nitrogen and hydrogen.

(3 marks)

(c)   The reaction is exothermic. Explain why a high temperature is still used.

(2 marks)

Making Esters

You'll probably remember <u>esters</u> from page 78... well, here's <u>a bit more</u> about them.

How to Make an Ester — **Reflux, Distil, Purify, Dry**

1) Making esters is a bit more complicated than just mixing an alcohol and a carboxylic acid together.

2) The reaction is <u>reversible</u> so some of the <u>ester</u> will react with the other product (<u>water</u>), and <u>re-form</u> the <u>carboxylic acid</u> and <u>alcohol</u>.

3) To get a <u>pure</u> ester you need a multi-step <u>reaction</u> and <u>purification</u> procedure.

1) **Refluxing** — The Reaction

To make ethyl ethanoate you need to react ethanol with ethanoic acid, using a <u>catalyst</u> such as <u>concentrated sulfuric acid</u> to speed things up.

<u>Heating</u> the mixture also speeds up the reaction — but you can't just stick a Bunsen under it as the ethanol will evaporate or catch fire before it can react.

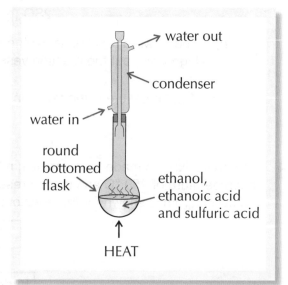

Instead, the mixture's <u>gently heated in a flask</u> fitted with a <u>condenser</u> — this catches the vapours and <u>recycles</u> them back into the flask, giving them <u>time to react</u>.

This handy method is called <u>refluxing</u>.

2) **Distillation**

The next step is <u>distillation</u>. This <u>separates</u> your lovely <u>ester</u> from all the other stuff left in the flask (<u>unreacted alcohol</u> and <u>carboxylic acid</u>, <u>sulfuric acid</u> and <u>water</u>).

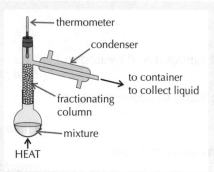

1) The mixture's <u>heated</u> below a fractionating column. As it starts to boil, the vapour goes up the <u>fractionating column</u>.

2) When the temperature at the top of the column reaches the <u>boiling point</u> of <u>ethyl ethanoate</u>, the liquid that flows out of the condenser is <u>collected</u>. This liquid is <u>impure</u> ethyl ethanoate.

Making Esters

3) Purification

The liquid collected (the <u>distillate</u>) is poured into a <u>tap funnel</u> and then treated to remove its impurities, as follows:

1) The mixture is shaken with <u>sodium carbonate solution</u> to remove acidic impurities.

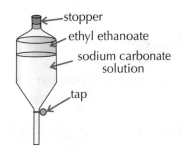

stopper
ethyl ethanoate
sodium carbonate solution
tap

2) Ethyl ethanoate doesn't mix with the water in the sodium carbonate solution, so the mixture <u>separates</u> into two layers, and the lower layer can be <u>tapped off</u> (removed).

stopper
ethyl ethanoate
calcium chloride solution
tap

3) The remaining upper layer is then shaken with <u>concentrated calcium chloride</u> solution to remove any <u>ethanol</u>. Again, the lower layer can be <u>tapped off</u> and removed.

4) Drying

1) Any remaining <u>water</u> in the ethyl ethanoate can be removed by shaking it with lumps of <u>anhydrous calcium chloride</u>, which absorb the water — this is called <u>drying</u>.

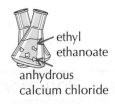

ethyl ethanoate
anhydrous calcium chloride

2) Finally, the <u>pure ethyl ethanoate</u> can be separated from the solid calcium chloride by <u>filtration</u>.

Esterification — it's 'esterical stuff...

So there you have it — the four steps that you need to make an ester. It's the refluxing the alcohol and acid bit where the ester's actually made — the other three steps are about separating it out from all the rest of the stuff in the reaction mixture. Then you get a nice pure ester at the end of it all.

Polymers

Plastics are made up of <u>lots</u> of molecules <u>joined together</u> in long chains.

Forces Between Molecules Determine The *Properties* of Plastics

<u>Strong covalent</u> bonds hold the <u>atoms</u> together in <u>long chains</u>. But it's the bonds <u>between</u> the different molecule chains that determine the <u>properties</u> of the plastic.

Weak Forces:

<u>Individual tangled chains</u> of polymers, held together by <u>weak intermolecular forces</u>, are free to <u>slide</u> over each other.

1) <u>THERMOSOFTENING POLYMERS</u> don't have cross-linking between chains.

2) The forces between the chains are really easy to overcome, so it's easy to <u>melt</u> the plastic.

3) When it <u>cools</u>, the polymer hardens into a new shape.

4) You can melt these plastics and <u>remould</u> them as many times as you like.

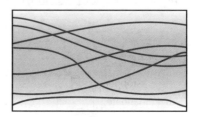

Strong Forces:

Some plastics have <u>stronger intermolecular forces</u> between the polymer chains, called <u>crosslinks</u>, that hold the chains firmly together.

1) <u>THERMOSETTING POLYMERS</u> have <u>crosslinks</u>.

2) These hold the chains together in a <u>solid structure</u>.

3) The polymer doesn't soften when it's heated.

4) Thermosetting polymers are <u>strong</u>, <u>hard</u> and <u>rigid</u>.

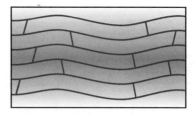

How You *Make* a *Polymer* Affects Its *Properties*

1) The <u>starting materials</u> and <u>reaction conditions</u> will both affect the properties of a polymer.

2) Two types of <u>polythene</u> can be made using different conditions:

- <u>Low density</u> (LD) polythene is made by heating ethene to about <u>200 °C</u> under <u>high pressure</u>. It's <u>flexible</u> and is used for bags and bottles.

- <u>High density</u> (HD) polythene is made at a <u>lower</u> temperature and pressure (with a catalyst). It's <u>more rigid</u> and is used for water tanks and drainpipes.

Polymers

There's plastic and there's... well, plastic. You wouldn't want to make a chair with the same plastic that gets used for flimsy old carrier bags. But whatever the plastic, it's always a polymer.

Polymers Can be **Modified** to Give Them **Different Properties**

You can chemically modify polymers to change their properties.

1) Polymers can be modified to increase their chain length.

   - Polymers with short chains are easy to shape and have lower melting points.
   - Longer chain polymers are stiffer and have higher melting points.

2) Polymers can be made stronger by adding cross-linking agents. These agents chemically bond the chains together, making the polymer stiffer, stronger and more heat-resistant.

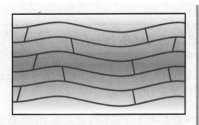

3) Plasticisers can be added to a polymer to make it softer and easier to shape. Plasticisers work by getting in between the polymer chains and reducing the forces between them.

4) The polymer can be made more crystalline. A crystalline polymer has straight chains with no branches so the chains can fit close together. Crystalline polymers have higher density, are stronger and have a higher melting point.

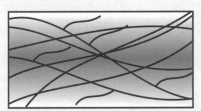

Branched polymer chains

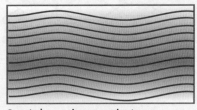

Straight polymer chains
— crystalline structure

Choose your polymers wisely...

The molecules that make up a polymer affect its properties and what it can be used for. If they're not useful enough, you can modify the polymers to make them even more useful.

New Materials

New materials are continually being developed, with new properties. These pages cover two different groups of (relatively) new materials — smart materials and nanoparticles.

Smart Materials Have Some Really Weird Properties

1) Smart materials behave differently depending on the conditions, e.g. temperature.

2) A good example is nitinol — a "shape memory alloy".
   It's a metal alloy (about half nickel, half titanium) but when it's cool you can bend it and twist it like rubber. Bend it too far, though, and it stays bent. But here's the really clever bit — if you heat it above a certain temperature, it goes back to a "remembered" shape.

3) It's really handy for glasses frames. If you accidentally bend them, you can just pop them into a bowl of hot water and they'll jump back into shape.

4) Nitinol is also used for dental braces. In the mouth it warms and tries to return to a 'remembered' shape, and so it gently pulls the teeth with it.

Nanoparticles are Really Really Really Really Tiny

...smaller than that.

1) Really tiny particles, 1–100 nanometres across, are called 'nanoparticles' (1 nm = 0.000 000 001 m).

2) Nanoparticles contain roughly a few hundred atoms.

3) Nanoparticles include fullerenes. These are molecules of carbon, shaped like hollow balls or closed tubes. The carbon atoms are arranged in hexagonal rings. Different fullerenes contain different numbers of carbon atoms.

4) A nanoparticle has very different properties from the 'bulk' chemical that it's made from — e.g. fullerenes have different properties from big lumps of carbon.

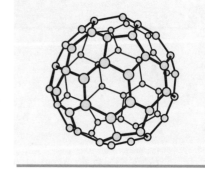

Fullerenes can Form Very Strong Nanotubes

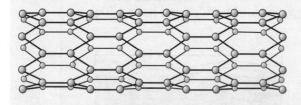

1) Fullerenes can be joined together to form nanotubes — teeny tiny hollow carbon tubes, a few nanometres across.

2) All those covalent bonds make carbon nanotubes very strong.

Nanomaterials are Often Designed for a Specific Use

Most nanomaterials are made using nanotechnology, but some nanoscale materials occur naturally or are produced by accident. For example:

- Seaspray — The sea produces nanoscale salt particles which are present in the atmosphere.
- Combustion — When fuels are burnt, nanoscale soot particles are produced.

Nanotechnology

Chemistry is so useful — who knew that applying sun tan cream was using such advanced technology.

Nanoparticles are Becoming More and More Widely Used

Using nanoparticles is known as <u>nanoscience</u>. Many <u>new uses</u> of nanoparticles are being developed.

1) They have a <u>huge surface area to volume ratio</u>, so they could help make new industrial <u>catalysts</u> (see page 169).

2) You can use nanoparticles to make <u>sensors</u> to detect one type of molecule and nothing else. These <u>highly specific</u> sensors are already being used to test water purity.

3) Nanotubes can be used to make <u>stronger</u>, <u>lighter</u> building materials.

4) New cosmetics, e.g. <u>sun tan cream</u> and <u>deodorant</u>, have been made using nanoparticles. The small particles do their job but don't leave <u>white marks</u> on the skin.

5) <u>Nanomedicine</u> is a hot topic. The idea is that tiny fullerenes are <u>absorbed</u> more easily by the body than most particles. This means they could <u>deliver drugs</u> right into the cells where they're needed.

6) New <u>lubricant coatings</u> are being developed using fullerenes. These coatings reduce friction a bit like <u>ball bearings</u> and could be used in all sorts of places from <u>artificial joints</u> to <u>gears</u>.

7) Nanotubes <u>conduct</u> electricity, so they can be used in tiny <u>electric circuits</u> for computer chips.

8) Nanoparticles are added to <u>plastics</u> in <u>sports equipment</u>, e.g. tennis rackets, golf clubs and golf balls. They make the plastic much <u>stronger</u> and <u>more durable</u>, and they don't <u>add weight</u>.

9) <u>Silver nanoparticles</u> are added to <u>polymer fibres</u> used to make <u>surgical masks</u> and <u>wound dressings</u>. This gives the fibres <u>antibacterial properties</u>.

Nanoparticles are much <u>smaller</u> than larger particles of the <u>same material</u>. This means they have a <u>larger surface area-to-volume ratio</u>, which is what gives them <u>different properties</u> and makes them super <u>useful</u>. <u>Silver nanoparticles</u>, for example, can <u>kill bacteria</u>, making them suitable for wound dressings. <u>Normal silver particles</u> are much bigger, have a smaller surface area to volume ratio and <u>can't</u> kill bacteria. Not so useful. Well, not for wound dressings anyway.

The Effects of Nanoparticles on Health are Not Fully Understood

1) Although nanoparticles are useful, the way they affect <u>the body</u> isn't fully understood, so it's important that any new products are <u>tested</u> thoroughly to minimise the risks.

2) Some people are worried that <u>products</u> containing nanoparticles have been made available <u>before</u> the effects on <u>human health</u> have been investigated <u>properly</u>, and that we don't know what the <u>long-term</u> impacts on health will be.

3) As the long-term impacts aren't known, many people believe that products containing nanoscale particles should be <u>clearly labelled</u>, so that consumers can choose whether or not to use them.

Bendy specs, tennis rackets and computer chips — cool...

Some nanoparticles have really <u>unexpected properties</u>. Silver's normally very unreactive, but silver nanoparticles can kill bacteria. Gold nanoparticles aren't gold-coloured — they're either red or purple. On the flipside, we also need to watch out for any unexpected harmful properties.

Warm-Up and Exam Questions

There are loads of practical uses for chemistry, particularly answering these and passing your exams.

Warm-Up Questions

1) Name the four stages used to make an ester.
2) What does a plastic's melting point tell you about the forces between its polymer chains?
3) What are intermolecular forces between polymer chains called?
4) Give two things that can affect the properties of a polymer.
5) What is nitinol?

Exam Questions

1 Scientists have developed new materials using nanoparticles, which show different properties from the same materials in bulk.

(a) Use words from the box to help you complete the sentences below.

| volume | mm | catalysts | surface area | nm | circuits |

(i) Nanoparticles are up to 100 .... in size.

(1 mark)

(ii) Nanoparticles have a large .... to

.... ratio.

(2 marks)

(b) Floyd Landis used a bike in the Tour de France with a frame weighing about 1 kg. Carbon nanotubes (CNT) were used in the manufacture of the frame of the bike.

(i) Suggest two properties of a material made using CNTs that make it suitable for use in a bike frame.

(2 marks)

(ii) Give the name of a type of molecule that can be joined together to make carbon nanotubes.

(1 mark)

(c) Give one other application of nanoparticles.

(1 mark)

2 The table below shows the properties of three polymers, **A**, **B** and **C**.

Give the polymer that would be best suited for each of the following uses:

(a) sandwich bag

(1 mark)

(b) drainpipe

(1 mark)

(c) disposable cup

(1 mark)

Polymer	Properties
A	heat resistant and strong
B	very flexible and biodegradable
C	very rigid

CFCs and the Ozone Layer

Scientists <u>changed their minds</u> about CFCs as they found out more about them.

At First Scientists Thought **CFCs were Great**...

1) <u>Chlorofluorocarbons</u> (CFCs for short) are <u>organic molecules</u> containing <u>carbon</u>, <u>chlorine</u> and <u>fluorine</u>, e.g. <u>dichlorodifluoromethane</u> CCl_2F_2 — this is like <u>methane</u> but with two chlorine and two fluorine atoms (and an extremely long name) instead of the four hydrogen atoms.

2) CFCs are <u>non-toxic</u>, <u>non-flammable</u> and chemically <u>inert</u> (unreactive). They're <u>insoluble</u> in water and have <u>low boiling points</u>. Scientists were very happy that they'd found some <u>non-toxic</u> and <u>inert</u> chemicals which were <u>ideal for many uses</u>.

3) Chlorofluorocarbons were used as <u>coolants</u> in <u>refrigerators</u> and <u>air-conditioning systems</u>.

4) CFCs were also used as <u>propellants</u> in <u>aerosol spray cans</u>.

...*But Then* They Discovered the **Shocking Truth**

This is called the "hole in the ozone layer".

1) In 1974 scientists found that <u>chlorine</u> could help to <u>destroy ozone</u> (see equations on p. 246).

2) In 1985 scientists found <u>evidence</u> of <u>decreasing ozone levels</u> in the atmosphere over Antarctica.

3) Measurements in the upper atmosphere show high levels of compounds produced by the <u>breakdown of CFCs</u>. This supports the hypothesis that CFCs break down and destroy ozone.

4) Scientists are now <u>sure</u> that CFCs are linked to the depletion (thinning) of the ozone layer.

- <u>Ozone</u> is a form of oxygen with the formula O_3 — it has <u>three oxygen atoms</u> per molecule, unlike ordinary oxygen which has <u>two</u> atoms per molecule.
- It hangs about in the <u>ozone layer</u>, way up in the <u>stratosphere</u> (part of the upper atmosphere), doing the very important job of <u>absorbing ultraviolet (UV) light</u> from the Sun. Ozone absorbs UV light and breaks down into an <u>oxygen molecule</u> and an <u>oxygen atom</u>: $O_3 + UV \text{ light} \rightarrow O + O_2$
- The oxygen molecule and oxygen atom join together to <u>make ozone again</u>: $O + O_2 \rightarrow O_3$
- <u>Reducing</u> the amount of <u>ozone</u> in the stratosphere results in <u>more UV light</u> passing through the atmosphere. Increased levels of <u>UV light</u> hitting the surface of the Earth can cause <u>medical problems</u> like increased risk of <u>sunburn</u> and <u>skin cancer</u>.

Some Countries Have *Banned* the Use of **CFCs**

1) Scientists' view that CFCs could damage the ozone layer caused a lot of <u>concern</u>.

2) But it took a while for society to do something about the mounting scientific evidence. Governments waited until the research had been thoroughly <u>peer reviewed</u> and <u>evaluated</u> before making a decision.

3) In 1978 the USA, Canada, Sweden and Norway <u>banned CFCs as aerosol propellants</u>.

4) After the <u>ozone hole</u> was discovered many countries (including the UK) got together and decided to reduce CFC production and eventually <u>ban CFCs completely</u>.

That's "chlorofluorocarbon" not "Chelsea Football Club"...

The ozone in the stratosphere is amazing stuff — it absorbs UV light and stops us from having to bear the full force of the Sun's UV output. Too much UV causes <u>sunburn</u> and <u>skin cancer</u>, so anything that damages the ozone and lets more UV through is a <u>bad</u> thing in the long run.

CFCs and the Ozone Layer

CFCs damage ozone by forming <u>free radicals</u>. Free radicals are made when <u>covalent bonds</u> split <u>evenly</u>...

Free Radicals are Made by Breaking Covalent Bonds

1) A <u>covalent bond</u>, remember, is one where <u>two atoms share electrons</u> between them, like in H_2 (page 117).

2) A covalent bond can <u>break unevenly</u> to form <u>two ions</u>, e.g. $\textbf{H–H} \rightarrow \textbf{H}^+ + \textbf{H}^-$. The H^- has <u>both</u> of the shared electrons, and the poor old H^+ has <u>neither</u> of them.

3) But a covalent bond can also break <u>evenly</u> — and then <u>each atom</u> gets <u>one</u> of the shared electrons, e.g. $\textbf{H–H} \rightarrow \textbf{H·} + \textbf{H·}$ — the H· is called a <u>free radical</u>. (The unpaired electron is shown by a <u>dot</u>.)

4) The unpaired electron makes the free radical <u>very, very reactive</u>.

Chlorine Free Radicals from CFCs Damage the Ozone Layer

1) <u>Ultraviolet light</u> makes the carbon-chlorine bonds in CFCs break up to form <u>free radicals</u>:

$$CCl_2F_2 \rightarrow CClF_2· + Cl·$$

Free radicals...

2) This happens <u>high up in the atmosphere</u> (in the <u>stratosphere</u>), where the <u>ultraviolet light</u> from the Sun is <u>stronger</u>.

3) <u>Chlorine free radicals</u> from this reaction react with <u>ozone</u> (O_3), turning it into ordinary oxygen molecules (O_2) and chlorine oxide (ClO·):

$$O_3 + Cl· \rightarrow ClO· + O_2$$

4) The chlorine oxide molecule is <u>very reactive</u>, and reacts with ozone to make two <u>oxygen molecules</u> and <u>another Cl· free radical</u>:

$$ClO· + O_3 \rightarrow 2O_2 + Cl·$$

5) This Cl· free radical now goes and reacts with <u>another ozone molecule</u>. This is a <u>chain reaction</u>, so just <u>one chlorine free radical</u> from one CFC molecule can go around breaking up <u>a lot of ozone molecules</u>.

CFCs <u>don't attack ozone directly</u>. They break up and form chlorine atoms (chlorine free radicals) which attack ozone. The chlorine atoms <u>aren't used up</u>, so they can carry on breaking down ozone.

CFCs and the Ozone Layer

CFCs have already been <u>banned</u> in many countries, but unfortunately they're <u>still destroying ozone</u>.

CFCs Stay in the Stratosphere for Ages

1) CFCs are <u>not very reactive</u> and will only react with one or two chemicals that are present in the atmosphere.

2) They only break up to form <u>chlorine atoms</u> in the stratosphere, where there's plenty of high-energy ultraviolet light around. They won't do it in the lower atmosphere.

3) This means that the CFCs in the stratosphere now will take a <u>long time</u> to be removed.

4) Remember, each CFC molecule produces one chlorine atom which can react with an <u>awful lot</u> of ozone molecules. <u>Thousands</u> of them, in fact.

5) So the millions of CFC molecules that are present in the stratosphere will continue to destroy ozone for a long time — even <u>after all CFCs have been banned</u>.

6) Each molecule will <u>stay around</u> for a long time, and each molecule will <u>destroy a lot of ozone</u> molecules.

Alkanes and HFCs are Safe Alternatives to CFCs

1) Alkanes <u>don't react</u> with ozone, so they can provide a safe alternative to CFCs.

2) <u>Hydrofluorocarbons</u> (<u>HFCs</u>) are compounds very similar to CFCs — but they contain <u>no chlorine</u>. It's the chlorine in CFCs that attacks ozone, remember.

3) Scientists have investigated the compounds that could be produced by breakdown of HFCs in the upper atmosphere, and <u>none of them</u> seem to be able to <u>attack ozone</u>. The <u>evidence suggests</u> that HFCs are <u>safe</u> to use.

The Montreal Protocol was an agreement to stop using CFCs

After discovering a <u>hole</u> in the <u>ozone layer</u>, many countries got together and decided to reduce CFC production and eventually <u>ban CFCs completely</u> — the agreement was called the <u>Montreal Protocol</u>.

Warm-Up and Exam Questions

Now that you've read all about CFCs and HFCs — long words with short abbreviations — try to get your head round these questions.

Warm-Up Questions

1) Give two reasons why CFCs were originally used in so many applications.

2) Give one use of CFCs.

3) What is a free radical?

4) Write an equation to show how CCl_2F_2 breaks up, forming a chlorine free radical.

5) Why will CFCs continue to deplete the ozone layer for many years after their use is banned?

Exam Questions

1   CFCs (chlorofluorocarbons) were used in many applications in the past, but it is now known that they have a destructive effect on the stratospheric ozone layer.

   (a)   Many CFC molecules only break up when they reach the stratosphere — why?

   (1 mark)

   (b)   One product of the break-up of CFCs is the chlorine free radical, Cl•.
         Complete the following two symbol equations showing its effect on ozone:

         (i) O_3 + Cl• → _____ + _____

         (ii)   _____ + O_3 → $2O_2$ + _____

   (4 marks)

   (c)   A small amount of Cl• can break up a lot of ozone molecules. Explain why.

   (2 marks)

   (d)   Name two groups of compounds that are now used as ozone-safe alternatives to CFCs.

   (2 marks)

2   (a)   Explain what 'the ozone layer' is.

   (2 marks)

   (b)   Name one region over which the Earth's ozone layer is thinner than it was 100 years ago.

   (1 mark)

   (c)   Discuss the possible connection between ozone levels and a rise in the incidence of skin cancer in humans.

   (3 marks)

3   Which of these statements about CFCs is false?

   A   CFCs break down to release chlorine, which reacts with ozone.
   B   CFCs are no longer used in aerosols because they are highly flammable.
   C   CFCs are organic molecules containing carbon, chlorine and fluorine.
   D   CFCs are very unreactive.

   (1 mark)

Revision Summary for Section 12

The end of another beautiful section and the end of the whole book — it brings a tear to my eye. Here's a handy pocket-sized checklist of things to make sure you've learnt: 1. definitions — learn what all the words mean, yes even the really long ones, 2. formulas — you'll lose easy marks on calculations if you don't, 3. examples — examiners love giving you marks for dropping the odd example in here and there, 4. reactions — these are the bread and butter of chemistry, 5. how to do things — diagrams often help here... I can't think of anything else right now, so try these questions to check I haven't forgotten anything vital.

1) What are 'fine chemicals'? Give an example.

2) What does LCA stand for?

3) What are the four stages that need to be considered when doing an LCA?

4) Explain why it is important to calculate how much of each reactant is needed to produce the desired amount of product before starting an industrial process.

5) Why is it important to choose the right apparatus for a chemical process?

6) What does calculating the yield tell you about a reaction?

7) Describe the five stages involved in producing chemicals in industry.

8) Give eight questions you should consider when deciding whether a process is sustainable.

9) What are the industrial conditions used for the Haber process?

10) What determines the choice of pressure used in the Haber process?

11) What effect does the iron catalyst have on the reaction between nitrogen and hydrogen?

12) Explain why nitrogen fixation is important to world food production and industry.

13) Write the symbol equations for the three reactions in the Contact Process.

14) State and explain the conditions used in the Contact Process.

15) Why is there a limit to the concentration of ethanol that can be made using fermentation? How can the concentration be increased? Sketch a diagram of the apparatus you could use.

16) Describe how ethanol is made from crude oil. What conditions are needed?

17) Make a table to compare the sustainability of the three methods of ethanol production (fermentation of sugar, fermentation of waste biomass, and from ethane).

18) You are making an ester. Explain why you would shake the distilled reaction mixture with:
   a) sodium carbonate.
   b) calcium chloride.

19) Explain the difference between thermosoftening and thermosetting polymers.

20) What would you add to a polymer to make it stiffer and stronger?

21) Give an example of a "smart" material and describe how it behaves.

22) What are nanoparticles? Name one type of nanoparticle.

23) Give three properties of CFC molecules.

24) What's the name for the part of the upper atmosphere that contains the ozone layer?

25) How are free radicals formed?

26) What are HFCs?

Page 18

Sulfur

Pages 20-21

Warm-Up Questions

1) The 'plum pudding model' — a positively charged 'pudding' containing negatively charged electrons.

2) protons and electrons

3) An element is a substance that consists of only one type of atom. There are about 100 elements.

4) Mass number is the sum of the number of protons and the number of neutrons in an atom. Atomic number is the number of protons (or electrons) in an atom.

5) a) 2

   b) 8

6) 2, 8, 3

7) molecules

Exam Questions

1 (a) 7 *(1 mark)*

   (b) N *(1 mark)*

   (c) non-metal *(1 mark)*

   (d) Any one of: e.g. phosphorus / arsenic / antimony / bismuth *(1 mark)*. It's in the same group as nitrogen so it will have the same number of electrons in its outer shell *(1 mark)* which gives it similar chemical properties *(1 mark)*.

2 (a) 19 *(1 mark)*

   (b) 19 *(1 mark)*

   (c) Number of neutrons = mass number – atomic number
       = 39 – 19 = **20** *(1 mark)*

   (d)

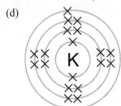

   (1 mark for 2 electrons in the first shell and a maximum of 8 in the other shells, 1 mark for correct number of electrons shown)

3 (a) 11 *(1 mark)*

   (b) Any one of: lithium / potassium / rubidium / caesium / francium *(1 mark)*

   (c) 1 *(1 mark)*

4 (a) The metal atoms lose electrons to form positive ions *(1 mark)* and the non-metal atoms gain electrons to form negative ions *(1 mark)*. The opposite charges (positive and negative) of the ions mean that they're strongly attracted to each other *(1 mark)*

   (b) In a non-metal compound each atom shares an electron with another atom *(1 mark)*. This is called covalent bonding *(1 mark)*.

5 B *(1 mark)*

6 C *(1 mark)*

7 How to grade your answer:

   0 marks: No description is given.

   1-2 marks: Brief description of how the theory of atomic structure has changed.

   3-4 marks: Some description of how the theory of atomic structure has changed. The answer has a logical structure and spelling, punctuation and grammar are mostly correct.

   5-6 marks: A detailed description of how the theory of atomic structure has changed. The answer has a logical structure and uses correct spelling, grammar and punctuation.

Here are some points your answer may include:

At the start of the 19th century John Dalton described atoms as solid spheres, made up of different elements.

In 1897 JJ Thomson concluded that atoms weren't solid spheres and that an atom must contain smaller, negatively charged particles — electrons. He called this the 'plum pudding model'.

In 1909 Ernest Rutherford conducted a gold foil experiment, firing positively charged particles at an extremely thin sheet of metal. Most of the particles went straight through, so they concluded that there was a positively charged nucleus at the centre, surrounded by a 'cloud' of negative electrons.

Niels Bohr proposed a new model of the atom where all the electrons were contained in shells. He suggested that electrons can only exist in fixed orbits, or shells, and that each shell has a fixed energy.

Pages 27-28

Warm-Up Questions

1) $H_{\diagdown O \diagup} H$

2) $C_6H_{12}O_6 + 6O_2 \rightarrow 6CO_2 + 6H_2O$

3) The $_2$ in H_2SO_4 refers to 2 atoms (of H), while the 2 in 2NaOH refers to 2 molecules of NaOH.

4) It either expands, or its pressure increases.

5) A solvent and a solute.

Exam Questions

1 (a) (i)   The particles are free to move past each other *(1 mark)* but there is some force of attraction between them so they tend to stick together *(1 mark)*.

       (ii) The particles are free to move / have virtually no force of attraction between them *(1 mark)* so they move in straight lines until they collide with each other or with the sides of the container *(1 mark)*.

       (iii) The particles have strong forces of attraction between them / are not free to move *(1 mark)* so they stay in a regular arrangement *(1 mark)*.

   (b) (i)   Particles at the surface of a liquid overcome the forces of attraction from other particles and escape *(1 mark)*.

       (ii) Perfumes have to evaporate easily so they can reach our smell receptors quickly *(1 mark)*.

2 C *(1 mark)*

3 (a) A solute is a substance that can be dissolved in a solvent *(1 mark)*.

   (b) Cyanoacrylate is insoluble in water, so water can't be used as the solvent *(1 mark)*. Cyanoacrylate is soluble in acetone, so acetone can be used as the solvent *(1 mark)*.

(a) sulfuric acid + ammonia → ammonium sulfate *(1 mark)*

(b) $H_2SO_4 + 2NH_3 \rightarrow (NH_4)_2SO_4$ *(1 mark for correct products and reactants, 1 mark for correctly balancing the equation)*

(c) 15 *(1 mark)*

There are eight atoms of hydrogen, one atom of sulfur, four atoms of oxygen, and two atoms of nitrogen.

(a) Nothing/it stays the same *(1 mark)*.

(b) 17 − 4 = 13 g
(1 mark for correct working, 1 mark for correct answer)

(c) (i) When you add the solute to water, the bonds holding the solute molecules together break and these molecules mix with the water molecules *(1 mark)*.

(ii) E.g. the molecules of substance Z are more strongly attracted to each other than to the water molecules *(1 mark)*. The water molecules are more strongly attracted to each other than to the molecules of substance Z *(1 mark)*.

Page 29
Revision Summary for Section 1

1) Calcium
2) 14 H and 6 C
3)
```
    H  H  H
    |  |  |
H—C—C—C—H
    |  |  |
    H  H  H
```

4) a) $CaCO_3 + 2HCl \rightarrow CaCl_2 + H_2O + CO_2$

b) $Ca + 2H_2O \rightarrow Ca(OH)_2 + H_2$

Page 36
Warm-Up Questions

Any two of, e.g. wood / cotton / wool / silk / leather

E.g. rubber / nylon / polyester / paint.

Limestone is used to make things like houses and roads. / Chemicals from limestone are used to make dyes, paints and medicines. / Limestone products are used to neutralise acidic soil/reduce acidity in lakes and rivers. / Limestone is used in power station chimneys to neutralise sulfur dioxide.

Limestone is widely available / it's cheaper than granite or marble / it's an easy rock to cut. / Limestone can be more hard-wearing than marble, but it still looks attractive. / Limestone can be used to produce concrete, which can be poured into moulds to make blocks or panels. This is a quick and cheap way of constructing buildings. / Limestone, concrete and cement don't rot when they get wet like wood does, / and they can't be gnawed away by insects. / Concrete doesn't corrode like a lot of metals do, / and it's fire resistant too.

evaporation

Exam Questions

(a) (i) A mixture of salt and impurities *(1 mark)*.

(ii) Water is injected into the salt deposit and dissolves the salt to make brine *(1 mark)*. Pressure forces the brine up to the surface *(1 mark)*. Impurities are removed from the brine in a refining plant and it's then pumped into containers *(1 mark)*. The brine is then boiled to make the water evaporate, leaving the salt behind *(1 mark)*.

(b) Seawater flows into specially built shallow pools *(1 mark)*. It is left to evaporate in the sun, leaving the salt behind *(1 mark)*. This process is repeated several times and then the salt is collected *(1 mark)*.

2 (a) $CaCO_3 \rightarrow CaO + CO_2$ *(1 mark)*

(b) A Bunsen burner can't reach a high enough temperature to thermally decompose all carbonates of Group 1 metals *(1 mark)*.

(c) Calcium carbonate in the limestone reacts with the acid in acid rain *(1 mark)* to form a calcium salt, carbon dioxide and water *(1 mark)*.

3 (a) Powdered limestone is heated in a kiln with powdered clay. *(1 mark for limestone and clay, 1 mark for heating in a kiln)*

(b) Limestone is made into cement and then mixed with sand and aggregate (water and gravel). *(1 mark for making into cement, 1 mark for sand and aggregate)*

(c) How to grade your answer:

0 marks: No negative impacts are given.

1-2 marks: Brief description of one or two of the negative impacts of quarrying limestone and using it to produce building materials.

3-4 marks: Description of at least three negative impacts of quarrying limestone and using it to produce building materials. The answer has a logical structure and spelling, punctuation and grammar are mostly correct.

5-6 marks: A detailed description of at least five of the negative impacts of quarrying limestone and using it to produce building materials is given. The answer has a logical structure and uses correct spelling, grammar and punctuation.

Here are some points your answer may include:

Quarrying makes big, unattractive holes in the landscape.

Quarrying permanently changes the landscape.

Quarrying processes, such as blasting rocks with explosives make lots of noise and dust in quiet, scenic areas.

Quarrying destroys the habitats of animals and birds.

Limestone is transported away from the quarry, usually in lorries, which causes noise and pollution.

Waste materials from quarries produce unsightly tips.

Quarried limestone can be used to make cement. Cement factories make a lot of dust, which can cause breathing problems for some people.

Energy is needed to produce cement. The energy is likely to come from burning fossil fuels, which causes pollution.

Pages 44-45
Warm-Up Questions

1) A rock which contains enough metal to make it worthwhile extracting the metal from it.

2) Any one of: e.g. zinc / iron / tin / copper.

3) It uses a lot of energy.

4) Phytomining involves growing plants in soil that contains copper. The plants can't use or get rid of the copper so it builds up in the leaves. The plants can be harvested, dried and burned. The copper is then collected from the ash.

5) E.g. it uses loads of energy / it scars the landscape / it destroys wildlife habitats.

Exam Questions

1 (a) It can be extracted by reduction with carbon *(1 mark)*, which is cheaper than electrolysis *(1 mark)*.

(b) The Cu^{2+} ions move towards the negative electrode/cathode *(1 mark)* and gain electrons to form copper atoms *(1 mark)*.

2 (a) E.g. the supply of copper-rich ores is limited / demand for copper is growing *(1 mark)*.

   (b) (i)   bioleaching *(1 mark)*

      (ii) Bacteria produce a (leachate) solution that contains copper *(1 mark)*.

      (iii) E.g. this method has a much smaller impact on the environment *(1 mark)*.

      (iv) E.g. it's slow *(1 mark)*.

3 How to grade your answer:

   0 marks: No positive or negative effects of mining metal ores are given.

   1-2 marks: Brief description of one positive and one negative effect of mining metal ores.

   3-4 marks: At least two positive and two negative effects of mining metal ores are given. The answer has a logical structure and spelling, grammar and punctuation are mostly correct.

   5-6 marks: The answer gives at least three positive and three negative effects of mining metal ores. The answer has a logical structure and uses correct spelling, grammar and punctuation.

   Here are some points your answer may include:

   Advantages:

   Mining metal ores allows useful products to be made from the metal.

   The mines provide jobs for workers.

   Mining brings money into the local area.

   Disadvantages:

   Mining can be very noisy.

   Mines scar the landscape.

   Mining can lead to a loss of wildlife habitats.

   Abandoned mine shafts can be dangerous.

4 (a) Any one of: e.g. potassium / sodium / calcium / magnesium / aluminium (any metal above carbon in the reactivity series) *(1 mark)*.

   (b) (i)   removal of oxygen / gain of electrons *(1 mark)*

      (ii) zinc oxide + carbon → zinc + carbon dioxide *(1 mark)*

   (c) $2Fe_2O_3 + 3C \rightarrow 4Fe + 3CO_2$ *(1 mark for the correct products and reactants, 1 mark for correctly balancing the equation)*

   (d) (i)   B *(1 mark)*

      (ii) Iron is more reactive than copper *(1 mark)*, so iron will displace copper *(1 mark)*.

   (e) E.g. extracting metals uses energy from fossil fuels *(1 mark)*. Recycling saves fossil fuels which are running out by saving energy *(1 mark)*. Using less fossil fuels means less pollution *(1 mark)*. Using less energy means recycling saves money *(1 mark)*. Recycling means less metals get sent to landfill sites *(1 mark)*. There's a finite amount of metal in the Earth so recycling will conserve these resources *(1 mark)*.

Pages 52-53

Warm-Up Questions

1) Any three of: e.g. strong / hard to break / can be bent or hammered into different shapes / conduct heat / conduct electricity.

2) In the centre block.

3) A mixture of metals, or a mixture of a metal and a non-metal.

4) alloys

5) iron + oxygen + water → hydrated iron(III) oxide

Exam Questions

1 (a) (i)   Impure iron straight from the blast furnace *(1 mark)*.

      (ii) Steel is twice as strong as cast iron *(1 mark)*.

      Cast iron has a tensile strength of 200 MPa, whereas steel has a tensile strength of 400 MPa.

   (b) Steel is relatively cheap and has a high tensile strength. Aluminium and copper are both more expensive and less strong than steel *(1 mark)*. Tungsten is stronger than steel but even more expensive than brass. Cast iron is cheaper than steel but it is a lot less strong *(1 mark)*.

   (c) (i)   It can be turned into an alloy by mixing it with other metals (or non-metals) *(1 mark)*.

      (ii) E.g. it's corrosion-resistant *(1 mark)*, it has a low density (it is light) *(1 mark)*.

2 (a) E.g. it doesn't corrode / it's not too bendy / it has a low density *(1 mark each, maximum 3 marks)*.

   (b) E.g. it conducts electricity / it conducts heat *(1 mark)*.

3 (a) E.g. high carbon steel is inflexible/very hard *(1 mark)*, but low carbon steel is easily shaped *(1 mark)*.

   (b) High carbon steel — e.g. blades for cutting tools / bridges *(1 mark)* Low carbon steel — e.g. car bodies *(1 mark)*

   (c) (i)   stainless steel *(1 mark)*

      (ii) e.g. cutlery / containers for corrosive substances *(1 mark)*

   (d) Scientists now know a lot about the properties of metals so alloys can be designed for specific uses *(1 mark)*.

4 (a) E.g. steel *(1 mark)*.

   (b) (i)   E.g. it has low density (is light) / it corrodes less than steel *(1 mark each)*.

      (ii) E.g. it is more expensive than steel *(1 mark)*.

   (c) E.g. to save natural resources *(1 mark)*, to save money *(1 mark)* and to reduce landfill use *(1 mark)*.

5 B *(1 mark)*

Page 59

Warm-Up Questions

1) Two (or more) elements or compounds that aren't chemically bonded together.

2) carbon and hydrogen only

3) Any three of, e.g. refinery gas (bottled gas) / petrol / naphtha / kerosene / diesel / oil / bitumen.

4) Hydrocarbons with more carbon atoms are more viscous.

5) Any three of: e.g. transport / electricity generation / making chemicals/ plastics / heating / covering roads (tarmac) / lubricating engine parts.

Exam Questions

1 (a) (i)   There should be an M in the bottom box *(1 mark)*.

      (ii) There should be a B in the top box *(1 mark)*.

      Fractions with longer molecules have a higher boiling point, so condense at the higher temperatures at the bottom of the column. Fractions with shorter molecules have a lower boiling point, so don't condense until they reach the top of the column.

   (b) The crude oil is heated and piped in at the bottom *(1 mark)*. The vaporised oil rises up the column *(1 mark)* and the fractions are tapped off at the different levels where they condense *(1 mark)*.

2 C *(1 mark)*

Pages 66-67

Warm-Up Questions

High temperature and a catalyst.

methane, ethane and propane

They contain carbon-carbon double bonds.

E.g. they're not broken down by microorganisms, so they don't rot.

complete combustion

Exam Questions

a) C_4H_{10} *(1 mark)*

b)

A *(1 mark)*

C *(1 mark)*

(a) Any two of: e.g. making LYCRA® fibre / window frames / synthetic leather / packaging / waterproof coating for fabrics / dental polymers / hydrogel wound dressings / memory foam *(1 mark each, up to 2 marks)*.

(b) E.g. most polymers aren't biodegradable, so it may be difficult/ expensive to dispose of waste. / Most polymers are made from crude oil, which is a non-renewable resource, so raw materials may become scarce/prices may rise *(1 mark)*.

(a)

(1 mark)

(b)
(1 mark)

(a) Complete combustion happens when there's plenty of oxygen available *(1 mark)*. Partial combustion happens when there's not enough oxygen available *(1 mark)*.

(b) B *(1 mark)*

(c) Partial combustion could produce carbon monoxide *(1 mark)*, which is a very toxic gas *(1 mark)*.

(a) Cracking is used to turn long-chain hydrocarbons, which aren't very useful, into smaller ones which are more useful *(1 mark)*.

(b) The long-chain hydrocarbons are heated to vaporise them *(1 mark)*. The vapour is passed over a powdered catalyst/aluminium oxide catalyst/mixed with steam *(1 mark)* at a high temperature/ 400 °C - 700 °C *(1 mark)*.

Page 68

Revision Summary for Section 3

) Propane — the fuel needs to be a gas at –10 °C to work in a camping stove.

Page 73

Warm-Up Questions

) They provide a lot of energy.

) A nickel catalyst, and a temperature of about 60 °C.

) They reduce the amount of cholesterol in the blood.

) An emulsion is a mixture of oil and water.

Exam Questions

1 (a) The plant material is crushed *(1 mark)*. The crushed plant material is pressed between metal plates to squash the oil out *(1 mark)*. The oil can be separated from the crushed plant material by a centrifuge or by using solvents *(1 mark)*. The oil is distilled to refine it *(1 mark)*.

(b) E.g. vegetable oils have higher boiling points than water so they can be used to cook foods at higher temperatures and at faster speeds *(1 mark)*. They give food a different flavour *(1 mark)*. They increase the amount of energy we get from food *(1 mark)*.

2 (a) hydrophilic ——●〰〰—— hydrophobic *(1 mark)*

(b) Hydrophobic means that part of the molecule is attracted to oil molecules *(1 mark)*. Hydrophilic means that part of the molecule is attracted to water molecules *(1 mark)*.

(c)

(1 mark)

(d) They prevent the mayonnaise emulsion separating into its component liquids / they keep the oil and water mixed well together. *(1 mark)*

Page 79

Warm-Up Questions

1) $C_nH_{2n+1}OH$

2) It's used as a cleaning fluid and as a fuel (other answers possible).

3) Methanoic acid, ethanoic acid and propanoic acid.

4) -COO-

Exam Questions

1 (a)

(1 mark for showing the –COOH functional group correctly, 1 mark for showing two carbons in total and correct hydrogens.)

(b) (i) ethyl ethanoate *(1 mark)*

(ii) water *(1 mark)*

(iii) Any one of: e.g. perfumes/flavouring/ointments/solvents *(1 mark)*.

(c) Carboxylic acids don't ionise completely in water *(1 mark)* so not many H^+ ions are released *(1 mark)*.

2 (a) E.g. methanol, propanol *(1 mark each)*

(b) -OH group *(1 mark)*

(c) $C_2H_5OH + 3O_2 \rightarrow 2CO_2 + 3H_2O$
(1 mark for correct reactants and products, 1 mark for correctly balancing equation)

(d) hydrogen *(1 mark)*

(e) Ethanol is flammable/it catches fire easily *(1 mark)*.

Pages 89-90

Warm-Up Questions

1) volcano

Earthquakes occur at plate boundaries but they aren't a geological feature.

2) A supercontinent made from the all present continents joined together.

3) The crust at the ocean floor is denser than the crust below the continents, so the denser oceanic plate will be forced below the less dense continental plate. Also, oceanic crust is cooler at the edges of the plate, so the edges will sink easily, pulling the oceanic plate down.

4) Igneous, metamorphic and sedimentary.

5) volcanic activity

6) fractional distillation

Exam Questions

1 (a) The diagram should be labelled:
A – crust *(1 mark)*
B – mantle *(1 mark)*
C – core *(1 mark)*

(b) (i) tectonic plates *(1 mark)*

(ii) Radioactive decay takes place in the mantle *(1 mark)*. This produces heat which causes the mantle to flow in convection currents *(1 mark)*. These currents cause the plates to drift *(1 mark)*.

2 (a) Any two of, the coastlines of South America and Africa fit together / very similar fossils have been found in rocks on different sides of the Atlantic, / certain rock layers of similar ages on different continents show similarity / there are living creatures found in both America and Africa that couldn't have crossed the Atlantic Ocean, e.g. earthworms. *(1 mark each)*.

(b) E.g. because people weren't convinced by his explanation of how the continents drifted. Wegener claimed the continents' movement could be caused by tidal forces and the Earth's rotation — but other geologists showed that this was impossible. *(1 mark)*

3 (a) From the action of heat and pressure *(1 mark)* over long periods of time *(1 mark)*.

(b) (i) limestone *(1 mark)*

(ii) Any two of, e.g. marble has smaller crystals / a more even texture / is harder *(1 mark each)*

(c) Sedimentary rocks form from layers of sediment laid down in lakes or seas *(1 mark)*. Over millions of years the layers get buried under more layers and the weight pressing down squeezes out the water *(1 mark)*. Also, fluids flowing through the pores deposit natural mineral cement between the sediment particles *(1 mark)*.

4 (a) Magma is molten rock beneath the Earth's surface *(1 mark)*. Lava is magma that's erupted from a volcano onto the Earth's surface *(1 mark)*.

(b) Igneous rocks contain different minerals in randomly arranged interlocking crystals, which makes them very hard *(1 mark)*.

(c) E.g. granite *(1 mark)*

5 A — 3 *(1 mark)*
B — 2 *(1 mark)*
C — 4 *(1 mark)*
D — 1 *(1 mark)*

6 (a) Billions of years ago, the Earth's atmosphere was rich in nitrogen, hydrogen, ammonia and methane *(1 mark)*. Lightning struck, causing a chemical reaction between the gases, resulting in the formation of amino acids *(1 mark)*. The amino acids collected in a 'primordial soup' *(1 mark)*. The amino acids gradually combined to produce organic matter, which eventually evolved into simple living organisms *(1 mark)*.

(b) In the 1950s, Miller and Urey sealed the gases that were present in the early atmosphere in their apparatus *(1 mark)*. They then heated them and applied an electrical charge for a week *(1 mark)*. They found that some amino acids had been made at the end of the experiment *(1 mark)*.

(c) 21% *(1 mark)*

7 Air is filtered to remove dust *(1 mark)*. It's then cooled to around –200 °C and becomes a liquid *(1 mark)*. During cooling water vapour condenses and is removed *(1 mark)*. Carbon dioxide freezes and is removed *(1 mark)*. The liquified air then enters the fractionating column and is heated slowly *(1 mark)*. The remaining gases are separated by fractional distillation *(1 mark)*.

Page 95
Warm-Up Questions

1) Acid Gas Scrubbers take the harmful gases out of the fumes produced by power stations before they are released into the atmosphere.

2) NO

3) When fossil fuels are burnt, nitrogen and oxygen in the air are exposed to very high temperatures and react to form nitrogen monoxide (NO). This nitrogen monoxide then reacts with more oxygen to form nitrogen dioxide (NO_2).

4) A fuel that won't run out.

5) E.g. ethanol, biodiesel, hydrogen gas.

Exam Questions

1 (a) global warming *(1 mark)*

(b) (i) Sulfur dioxide produced from burning fossil fuels mixes with clouds *(1 mark)* and forms dilute sulfuric acid *(1 mark)*. This then falls as acid rain *(1 mark)*.

(ii) Any one of: e.g. makes lakes acidic and many plants and animals die as a result / kills trees / damages limestone buildings and stone statues / could possibly cause health problems *(1 mark)*.

(c) (i) Global dimming is the reduction in the amount of sunlight reaching the Earth's surface *(1 mark)*.

(ii) Global dimming is caused by particles of soot and ash produced when fossil fuels are burnt that reflect sunlight back into space *(1 mark)*.

(d) Any two of: e.g. engines need to be converted to work with ethanol. / Ethanol is not widely available. / It may increase food prices if farmers switch to growing crops to make ethanol from growing crops to make food *(1 mark each, up to 2 marks)*.

2 (a) vegetable oils *(1 mark)*

(b) (i) Global warming is caused by an increase in carbon dioxide *(1 mark)*. Biodiesel is 'carbon neutral' *(1 mark)*.

(ii) Any two of: e.g. produces less sulfur dioxide than diesel or petrol / doesn't release as many particulates as diesel or petrol / engines don't need to be converted *(1 mark each, up to 2 marks)*.

(c) Any two of: e.g. it's expensive to make / we're unable to make enough biodiesel to completely replace regular diesel / it could increase food prices if farmers switch from growing food crops to growing crops to make biodiesel *(1 mark each, up to 2 marks)*.

3 (a) It's made from the electrolysis of water *(1 mark)*.

(b) It's a clean fuel — it combines with oxygen to form just water *(1 mark)*.

(c) Any one of: e.g. you would need a special expensive engine / hydrogen isn't widely available / you still need energy from another source to make hydrogen / hydrogen is hard to store *(1 mark)*.

Page 96
Revision Summary for Section 5

2) b) 2 cm
c) 3.5 years

Page 101

Warm-Up Questions

* The groups of three that Döbereiner organised the elements into, according to their chemical properties.
* Mendeleev left gaps in his table so that elements with similar properties would be in the same vertical groups.
* E.g. the number of electrons in its highest occupied energy level.
* isotopes

Exam Questions

(a) An isotope is a different atomic form of the same element *(1 mark)*, which has the same number of protons *(1 mark)* but a different number of neutrons *(1 mark)*.

(b)

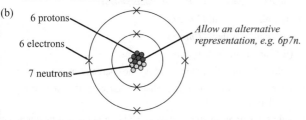

6 protons
6 electrons
7 neutrons

Allow an alternative representation, e.g. 6p7n.

(1 mark each for the correct number and placement of protons, neutrons and electrons)

(a) Relative atomic mass *(1 mark)*.

(b) Boron-11 has one more neutron in its nucleus than boron-10 *(1 mark)*.

(c) Boron-11 must be the most abundant *(1 mark)*. The A_r value takes into account the relative abundance of each isotope, and in the case of boron it is closer to 11 than to 10 *(1 mark)*.

D *(1 mark)*

(a) Newlands listed elements in rows of seven in order of their relative atomic mass *(1 mark)*.

(b) E.g. Newlands' groups contained elements that didn't have similar properties, e.g. carbon and titanium *(1 mark)*. He mixed up metals and non-metals, e.g. oxygen and iron *(1 mark)*. He didn't leave any gaps for elements that hadn't been discovered yet unlike Mendeleev *(1 mark)*. This meant that Newlands' Octaves broke down on the third row *(1 mark)*.

Pages 107-108

Warm-Up Questions

1) A high boiling point.
2) When ionic compounds are dissolved the ions separate and are free to move in the solution. These free-moving charged particles allow the solution to carry electric current.
3) positive ions
4) negative ions
5) $Al(OH)_3$

Exam Questions

(a) Sodium will lose the electron from its outer shell to form a positive ion *(1 mark)*. Fluorine will gain an electron to form a negative ion *(1 mark)*.

(b) Ionic compounds have a giant ionic lattice structure *(1 mark)*. The ions form a closely packed regular lattice arrangement *(1 mark)*, held together by strong electrostatic forces of attraction between oppositely charged ions *(1 mark)*.

(a) $MgCO_3$ *(1 mark)*

(b) Li_2SO_4 *(1 mark)*

---

3 (a) lithium oxide *(1 mark)*

(b) (i) and (ii)

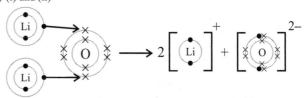

(1 mark for arrows shown correctly, 1 mark for correct electron arrangement and charge on lithium ion, 1 mark for correct electron arrangement and charge on oxygen ion).

4 (a)

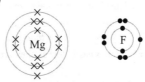

(1 mark)   *(1 mark)*

(b) Mg^{2+} *(1 mark)* and F^- *(1 mark)*

(c) MgF_2 *(1 mark)*

(d) There are electrostatic forces of attraction between the ions *(1 mark)*.

(e) (i) There are strong electrostatic forces between the ions *(1 mark)* so a large amount of energy is needed to break these bonds/overcome these forces *(1 mark)*.

(ii) When the magnesium fluoride is molten the ions can move about and carry charge (i.e. conduct a current) through the liquid *(1 mark)*.

5 (a)

	Potassium atom, K	Potassium ion, K^+	Chlorine atom, Cl	Chloride ion, Cl^-
Number of electrons	19	18	17	18

(2 marks for all three columns correct, otherwise 1 mark for any two columns correct)

(b)

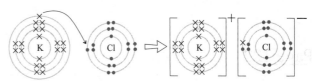

(2 marks — 1 mark for correct electron arrangements, 1 mark for correct arrow and charges on ions)

Pages 115-116

Warm-Up Questions

1) it increases
2) hydrogen
3) The elements go from gas to liquid to solid (at room temperature).
4) E.g. NaCl / KBr / LiI

You can have any Group 1 with Group 7 salt here (there are loads) and your answer can be a formula (e.g. NaCl) or in words (e.g. sodium chloride).

5) $Br_{2(aq)} + 2KI_{(aq)} \rightarrow I_{2(aq)} + 2KBr_{(aq)}$
6) As noble gases don't react with much, they are colourless and non-flammable, they are hard to observe.

Exam Questions

1 (a)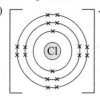

(1 mark)

(b) (i) chlorine + potassium bromide → bromine + potassium chloride *(1 mark)*

(ii) orange/brown *(1 mark)*

(iii) It's easier for chlorine to gain electrons *(1 mark)* because the outer shell is closer to the nucleus than in bromine *(1 mark)*.

Atoms react by gaining or losing electrons. Bromine is less able to attract an extra electron as its outer shell is further from the nucleus than chlorine's.

2 (a) They have a single outer electron which is easily lost so they are very reactive *(1 mark)*.

(b) As you go down Group 1, the outer electron is further from the nucleus *(1 mark)*. So it is more easily lost and the alkali metal is more reactive *(1 mark)*.

(c) hydrogen *(1 mark)* and a hydroxide *(1 mark)*

(d) alkaline *(1 mark)*

3 (a) Fluorine — gas *(1 mark)*

Chlorine — gas *(1 mark)*

Bromine — liquid *(1 mark)*

Iodine — solid *(1 mark)*

(b) Arrow should be pointing upwards. (1 mark)

(c) (i) displacement *(1 mark)*

Chlorine is displacing iodine.

(ii) iodine / I / I_2 *(1 mark)*

4 (a) Group 0 *(1 mark)*

(b) As you move down the group, the boiling points of the noble gases increase *(1 mark)*.

Page 122

Example

A is simple molecular, B is giant metallic, C is giant covalent, D is giant ionic

Pages 123-124

Warm-Up Questions

1) In a covalent bond, the atoms share electrons. In an ionic bond, one of the atoms donates electrons to the other atom.

2) Because the intermolecular forces between the chlorine molecules are very weak.

3) Any two of: e.g. diamond is very hard and graphite is fairly soft. / Graphite conducts electricity and diamond doesn't. / Diamond is clear/ transparent and graphite is opaque. / Diamond is colourless and graphite is black.

4) E.g. silicon dioxide/silica

Exam Questions

1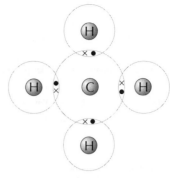

(1 mark for bonds shown correctly, 1 mark for correct number of atoms shown)

2 (a) (i) giant covalent *(1 mark)*

(ii) giant covalent *(1 mark)*

(iii) simple molecular *(1 mark)*

(b) It doesn't contain any ions to carry the charge *(1 mark)*.

(c) Each carbon atom has a delocalised electron that's able to carry the charge *(1 mark)*.

(d) All of the atoms in silicon dioxide and in graphite are held together by strong covalent bonds *(1 mark)*. In bromine, each molecule is held together with a strong covalent bond but the forces between these molecules are weak *(1 mark)*.

In order to melt, a substance has to overcome the forces holding its particles tightly together in the rigid structure of a solid. If the forces between the particles are weak, this is easy to do and doesn't take much energy at all. But if the forces are really strong, like in a giant covalent structure, you have to provide loads of heat to give the particles enough energy to break free.

3 (a) E.g:

(1 mark for showing two different sizes of atoms, 1 mark for showing irregular arrangement)

(b) The regular arrangement of atoms in iron means that they can slide over each other meaning iron can be bent *(1 mark)*. Steel contains different sized atoms which distorts the layers of iron atoms *(1 mark)*, making it more difficult for them to slide over each other *(1 mark)*.

4 (a) solid *(1 mark)*

(b) (i) giant covalent *(1 mark)*

(ii) giant ionic *(1 mark)*

(c) (i)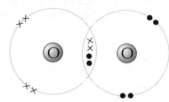

(1 mark for correct number of electrons in each shell, 1 mark for showing covalent bond correctly)

(ii) Oxygen has weak intermolecular forces between its molecules *(1 mark)*. It doesn't take much energy to separate the molecule *(1 mark)*.

(d) (i) It has delocalised electrons *(1 mark)* which are free to move through the whole structure and carry an electrical charge *(1 mark)*.

(ii) yes *(1 mark)*

Page 129

Warm-Up Questions

) E.g. it's very expensive to make the metal cold enough. / It's difficult to make the metal cold enough.

) Any three of, e.g. more than one ion (with different charges) / coloured compounds / make useful catalysts / good conductors of heat/electricity / dense / strong / shiny / hard / have high melting points.

) thermal decomposition

) precipitation

Exam Questions

(a) Transition metals have higher melting points than group 1 and 2 metals / transition metals have higher densities than group 1 and 2 metals / transition metals can form ions with different charges / transition metals react less vigorously when heated in air than group 1 and 2 metals *(3 marks available — 1 mark for each correct point)*.

Questions like this are easy marks — all you're asked to do is find information from the table — the answers are right there in front of you.

(b) E.g. catalytic properties (as metals or compounds) *(1 mark)*, transition metal compounds are usually coloured *(1 mark)*.

(c) (i) Bubble the gas through limewater — it will turn cloudy if carbon dioxide is present *(1 mark)*.

(ii) iron(II) carbonate → iron oxide + carbon dioxide *(1 mark)*.

(a) 'Superconducting' means that the substance has no electrical resistance at all when a current flows through it *(1 mark)*, meaning that none of the electrical energy is wasted as heat *(1 mark)*.

(b) Metal oxide ceramic 2, because it's easier to cool substances to this temperature and keep them there than it is to reach lower temperatures *(1 mark)*.

In fact, none of these substances are likely to be a lot of use in the real world. Keeping an electromagnet or power line at this temperature all the time would be really hard to do. Room temperature (or close to it) superconductors are what's needed, which may sound impossible — but think of all the stuff science has come up with that must have seemed impossible once...

(c) Any two of, power cables that carry electricity without losing any power / very fast electrical circuits / very strong electromagnets that can work without a constant power source *(1 mark each)*

Page 130

Revision Summary for Section 6

) 20.18

)A: giant metallic,
   B: giant covalent,
   C: giant ionic

Page 132

) 30.0%

) 88.9%

) 48.0%

) 65.3%

Pages 136-137

Warm-Up Questions

1) The relative formula mass.

2) A mole is the relative formula mass of a substance, in grams.

3) One mole of O_2 weighs $16 \times 2 = 32$ g.

4) 0.1 mole

Exam Questions

1 (a) M_r of $BF_3 = 11 + (19 \times 3) = 68$ *(1 mark)*

(b) M_r of $B(OH)_3 = 11 + (17 \times 3) = 62$ *(1 mark)*

2 (a) $100 - 60 = 40\%$ *(1 mark)*

(b) 40 g of sulfur combine with 60 g of oxygen.

S = 40	O = 60
40 ÷ 32	60 ÷ 16
= 1.25	= 3.75
1.25 ÷ 1.25 = 1	3.75 ÷ 1.25 = 3

Therefore, the formula of the oxide is SO_3
(2 marks — 1 mark for correct working)

(c) 2460 cm$^3 = 2.46$ dm^3
Number of moles = volume of gas ÷ volume of 1 mole
= 2.46 ÷ 24
= 0.103 moles
(2 marks — 1 mark for correct working)

3 (a) 100g .... reacts to give ... 56 g
1 g .... reacts to give ... 56 ÷ 100 = 0.56 g
2 g .... reacts to give ... 0.56 × 2 = 1.12 g *(1 mark)*

(b) E.g. When transferring the $CaCO_3$ from the weighing apparatus to the test tube, or the CaO from the test tube to the weighing apparatus some of the solid may be left behind *(1 mark)*.

4 (a) % mass of N in $CO(NH_2)_2 = [(A_r \times$ no. of atoms) ÷ M_r] × 100
= [(14 × 2) ÷ (12 + 16 + 32)] × 100
= 47%
(1 mark for correct working, 1 mark for correct answer)
% mass of N in KNO_3 = [14 ÷ (39 + 14 + 48)] × 100
= 14%
(1 mark for correct working, 1 mark for correct answer)
% mass of N in NH_4NO_3 = [(14 × 2) ÷ (28 + 4 + 48)] × 100
= 35%
(1 mark for correct working, 1 mark for correct answer)

Calculations are often worth more than one mark. It can be tempting just to scribble down enough working out to get you to the answer. But it's worth bearing in mind that if you get the final answer wrong, you could still get some marks for the working. So if you put down each step clearly it could pay off.

(b) Urea *(1 mark)*. It contains the greatest percentage mass of nitrogen, so would provide more nitrogen for plant growth per kg spread on the soil *(1 mark)*.

5 (a) Relative formula mass = 23 + 16 + 1 = 40 *(1 mark)*

(b) number of moles = mass (g) ÷ M_r
= 4 ÷ 40 = **0.1 moles**
(1 mark for correct working, 1 mark for correct answer)

(c) 240 g *(1 mark)*. In a chemical reaction, mass is always conserved, so if there are 246 g of reactants, there will be 246 g of products in total *(1 mark)*.

6 (a) Proportion of C in CO_2 = 12 ÷ 44 ≈ 0.27
mass of C in compound = 4.4 × 0.27 ≈ 1.2 g *(1 mark)*
moles of C = 1.2 ÷ 12 = 0.1 mol *(1 mark)*

(b) Proportion of H in H_2O = 2 ÷ 18 ≈ 0.11
mass of H in compound = 1.8 × 0.11 ≈ 0.2 g *(1 mark)*
moles of H = 0.2 ÷ 1 = 0.2 mol *(1 mark)*

(c) Ratio of C:H = 0.1:0.2 = 1:2
So the empirical formula is CH_2 *(1 mark)*.

7 60 g of calcium combine with 106.5 g of chlorine.

Ca = 40   Cl = 35.5

60 ÷ 40   106.5 ÷ 35.5

= 1.5   = 3

1.5 ÷ 1.5 = 1   3 ÷ 1.5 = 2

Therefore, the formula of calcium chloride is $CaCl_2$
(2 marks — 1 mark for correct working)

Page 142

Warm-Up Questions

1) Waste by-products decrease the atom economy of a reaction.

2) 100%.

All reactions with one product will have 100% atom economy.

3) Because they use up resources very quickly and produce a lot of waste. This has to be disposed of (e.g. in landfill sites or in the sea).

4) Percentage yield = (4 ÷ 5) × 100
= 80%

5) Because a low product yield means that resources are wasted rather than being saved for future generations.

Exam Questions

1 M_r of ethanol = (12 × 2) + 6 + 16 = 46
M_r of ethene = (12 × 2) + 4 = 28 *(1 mark)*
Atom economy = (28 ÷ 46) × 100 *(1 mark)*
= 61 % *(1 mark)*

2 (a) From the equation, 4 moles of CuO → 4 moles of Cu
so 1 mole CuO → 1 mole Cu *(1 mark)*

63.5 + 16 = 79.5 g CuO → 63.5 g Cu *(1 mark)*

1 g CuO → 63.5 ÷ 79.5 = 0.8 g (1 d.p.)

4 g CuO → 0.8 × 4 = 3.2 g *(1 mark)*

(b) Percentage yield = (2.8 ÷ 3.2) × 100 *(1 mark)*
= 87.5% *(1 mark)*

(c) Any three of: e.g

There may have been an incomplete reaction — some copper oxide was not reduced *(1 mark)*.

There may have been unexpected reactions (which produced different products) due to impurities in the reactants *(1 mark)*.

Some of the copper may have been left behind when it was scraped out into the beaker *(1 mark)*.

Some of the copper may have been left on the filter paper *(1 mark)*.

3 How to grade your answer:

0 marks: No reasons why yields are always less than 100% are given.

1-2 marks: Brief description of one reason why yields are less than 100% is given.

3-4 marks: Two or three reasons why yields are always less than 100% are given. The answer has a logical structure and spelling, grammar and punctuation are mostly correct.

5-6 marks: Four or five reasons why yields are always less than 100% are given. The answer has a logical structure and uses correct spelling, grammar and punctuation.

Here are some points your answer may include:

Some reactions are reversible. Some of the reactants will never be completely converted to products because the reaction goes both ways.

If a liquid is filtered to remove solid particles, some of the liquid or solid could be lost when it's separated from the reaction mixture.

Transferring solutions from one container to another often leaves behind traces on the containers.

Liquids evaporate all the time, and evaporation is greater if they're heated.

There may be unexpected reactions happening. These reaction will use up reactants, so there's not as much to make the desired product.

4 D *(1 mark)*

Page 143

Revision Summary for Section 7

2 (a) 40   (b) 108   (c) 44   (d) 84
(e) 106   (f) 81   (g) 56   (h) 17

4 (a) (i) 12.0%   (ii) 27.3%   (iii) 75.0%
(b) (i) 74.2%   (ii) 70.0%   (iii) 52.9%

5 (b) $MgSO_4$

7 80.3 g

Pages 150-151

Warm-Up Questions

1) Oxidising — provides oxygen which allows other materials to burn more fiercely.

2) neutralisation

3) Strong acids ionise completely in water so every hydrogen atom releases a hydrogen ion. Weak acids do not fully ionise, so only some of the hydrogen atoms in the compound release hydrogen ions.

4) copper nitrate and water

5) calcium sulfate, water and carbon dioxide

Exam Questions

1 (a) A — red
B — pH 7
D — pH 8/9
E — purple
(2 marks if all correct, 1 mark for 2 or 3 correct)

(b) C *(1 mark)*

(c) E *(1 mark)*

(d) B *(1 mark)*

(e) A *(1 mark)*

With questions like this, always have a guess if you're not sure. Remember, the examiners can't take marks off you (even for a really silly answer) and if you're stuck between two possibilities you're much more likely to get a mark if you go for one of them than if you put nothing at all.

(a) alkalis *(1 mark)*

(b) (i) MgO + 2HCl → MgCl$_2$ *(1 mark)* + H$_2$O *(1 mark)*

   (ii) magnesium oxide + hydrochloric acid → magnesium chloride + water *(1 mark)*

(c) carbon dioxide *(1 mark)*

(d) 2HNO$_3$ + CuCO$_3$ → Cu(NO$_3$)$_2$ + H$_2$O + CO$_2$
(1 mark for correct reactant and products, 1 mark for correctly balancing the equation)

(a) acid + metal → salt + hydrogen *(1 mark)*

(b) The bubbles are caused by the hydrogen produced in the reaction *(1 mark)*. The speed of reaction is indicated by the rate at which the bubbles of hydrogen are given off *(1 mark)*. Magnesium is more reactive, so it reacts with the acid producing more bubbles *(1 mark)*.

(c) (i) There will be a squeaky pop *(1 mark)*.

   (ii) There will not be a squeaky pop *(1 mark)*. Copper does not react at all with acids because it is less reactive than hydrogen *(1 mark)*.

(d) H$_2$SO$_4$ + Mg → MgSO$_4$ + H$_2$ *(1 mark)*

Acid A as the electrical conduction reading is higher *(1 mark)*. Strong acids ionise completely in water, so lots of H$^+$ ions are released *(1 mark)*. Weak acids do not fully ionise in water, so fewer H$^+$ ions are released *(1 mark)*. Ions carry electric charge through acid solutions *(1 mark)*. This means that strong acids are better conductors of electricity *(1 mark)*.

Page 157

Warm-Up Questions

) soluble

) barium chloride + sodium sulfate → barium sulfate + sodium chloride

) Add the insoluble base to an acid until all the acid is neutralised and the excess base can be seen on the bottom of the flask. Then filter out the excess base and evaporate off the water to leave a pure, dry sample.

) One substance is oxidised and loses electrons. At the same time, another is reduced and gains electrons.

) hydrated iron(III) oxide

) tin

) magnesium

Exam Questions

(a) It must be insoluble *(1 mark)*.

(b) silver nitrate + hydrochloric acid → silver chloride + nitric acid *(1 mark)*.

(c) First, filter the solution to remove the salt which has precipitated out *(1 mark)*. Then wash the insoluble salt *(1 mark)* and then leave it to dry on filter paper *(1 mark)*.

(a) iron + water + oxygen → hydrated iron(III) oxide *(1 mark)*

(b) Iron loses electrons *(1 mark)* when it reacts with oxygen and is oxidised *(1 mark)*. Oxygen gains electrons *(1 mark)* when it reacts with iron and is reduced *(1 mark)*.

(c) The coat of zinc acts as sacrificial protection — it's more reactive than iron so it'll lose electrons in preference to iron *(1 mark)*. The zinc also acts as a barrier to oxygen and water *(1 mark)*.

(d) The coating of tin acts as a barrier, stopping water and oxygen in the air from reaching the surface of the iron *(1 mark)*.

Page 166

Warm-Up Questions

1) E.g. the corrosion of iron is a reaction that happens very slowly. Explosions are very fast reactions.

2) Any three of, e.g. Increase the temperature (of the acid). / Use smaller pieces of/powdered magnesium. / Increase the acid concentration. / Use a catalyst.

3) It would increase the time taken (i.e. reduce the rate of reaction).

4) E.g. by keeping the milk cool/storing it in a fridge.

5) E.g. measure the volume of gas given off by collecting it in a gas syringe / monitor the mass of a reaction flask from which the gas escapes.

Exam Questions

1 (a) Any two from: e.g. the concentration of sodium thiosulfate/ hydrochloric acid / the person judging when the black cross is obscured / the black cross used (size, darkness etc.) *(1 mark each)*.

Judging when a cross is completely obscured is quite subjective — two people might not agree on exactly when it happens. You can try to limit this problem by using the same person each time, but you can't remove the problem completely. The person might have changed their mind slightly by the time they do the next experiment — or be looking at it from a different angle, be a bit more bored, etc.

(b)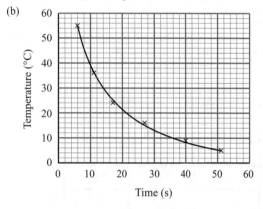

(1 mark for all points plotted correctly, 1 mark for best-fit curve)

(c) As the temperature increases the time decreases, meaning that the reaction is happening faster *(1 mark)*.

(d) Each of the reactions would happen more slowly *(1 mark)*, although they would still vary with temperature in the same way *(1 mark)*.

(e) E.g. by repeating the experiment and taking an average of the results *(1 mark)*.

2 (a) A gas/carbon dioxide is produced and leaves the flask *(1 mark)*.

(b) The same volume and concentration of acid was used each time, with excess marble *(1 mark)*.

(c) E.g. the marble chips were smaller/the temperature was higher *(1 mark)*.

(d) The concentration of the acid is greatest at this point, before it starts being converted into products *(1 mark)*.

Page 172

Warm-Up Questions

1) They must collide with enough energy.

2) There will be more frequent collisions so the rate of reaction will increase.

3) A catalyst is a substance which speeds up a reaction, without being chemically changed or used up in the reaction.

4) The temperature will decrease.

260

Exam Questions

1   How to grade your answer:

   0 marks: No ways of increasing the rate given.

   1-2 marks: One or two ways of increasing the rate are given.
                      No discussion of collision theory is provided.

   3-4 marks: At least two ways of increasing the rate are given.
                      There is some relevant discussion of collision theory.
                      The answer has a logical structure and spelling,
                      grammar and punctuation are mostly correct.

   5-6 marks: Detailed discussion of ways of increasing the rate and
                      relevant collision theory is given. The answer has a
                      logical structure and uses correct spelling, grammar
                      and punctuation.

   Here are some points your answer may include:

   Collision theory says that the rate of reaction depends on how often
   and how hard the reacting particles collide with each other. If the
   particles collide hard enough (with enough energy) they will react.

   Increasing the temperature makes particles move faster, so they
   collide more often and with greater energy. This will increase the
   rate of reaction.

   If the surface area of the catalyst is increased then the particles
   around it will have more area to work on. This increases the
   frequency of successful collisions and will increase the rate
   of reaction.

   Increasing the pressure of the hydrogen will mean the particles are
   more squashed up together. This will increase the frequency of the
   collisions and increase the rate of reaction.

2   (a) Neutralisation / Exothermic *(1 mark)*.

   (b)

Time (s)	Temperature of the reaction mixture (°C)		
	1st run	2nd run	Average
0	22	22	22.0
1	25.6	24.4	25.0
2	28.3	28.1	28.2
3	29.0	28.6	28.8
4	28.8	28.8	28.8
5	28.3	28.7	28.5

   (2 marks if all correct, 1 mark for 4 or 5 correct)

   (c) (28.8 – 22.0 =) 6.8 °C *(1 mark)*

   *Don't get caught off guard — the maximum average change is just the
   highest average temperature minus the lowest average temperature.*

   (d) Exothermic *(1 mark)* because heat is given out to the surroundings/
   the temperature of the reaction mixture increases *(1 mark)*.

Pages 181-182

Warm-Up Questions

1) formed

2) ΔH

3) A catalyst lowers the activation energy (by providing a different pathway
   for the reaction).

4) The equilibrium will move to the right/towards products.

5) The equilibrium will move to the left/towards reactants.

6) It would have no effect (because there are equal numbers of gas
   molecules on both sides).

Exam Questions

1   (a) 20% *(1 mark)* (Accept 18–22%)

   (b) As the pressure increases, the equilibrium moves to the left/towards
      the reactants *(1 mark)*, because there are fewer gas moles on the left,
      more gas moles on the right *(1 mark)* and the equilibrium shifts to
      oppose the change *(1 mark)*.

   (c) (i)   As the temperature increases the yield increases *(1 mark)*, so the
           equilibrium has moved to the right/towards products *(1 mark)*.

      (ii) endothermic *(1 mark)*

2   (a) C—H and O = O *(1 mark)*

   (b) (i)   (4 × 414) + (2 × 494) = 2644 kJ/mol *(1 mark)*

      (ii) (2 × 800) + (4 × 459) = 3436 kJ/mol *(1 mark)*

      (iii) 3436 – 2644 = 792 kJ/mol *(1 mark)*

   (c) The energy released when the new bonds are formed is greater than
      the energy needed to break the original bonds, so overall energy is
      given out *(1 mark)*.

3   (a) E.g. so that the temperature rise is proportional to the amount of heat
      produced *(1 mark)*.

   (b) 100 × 4.2 × 21 *(1 mark)* = 8820 J/8.82 kJ *(1 mark — units needed)*

   *This gives you the energy transferred (in J) and normally you would then
   have to divide this by the mass of fuel burned (in g) to find the heat
   energy transferred per gram of fuel. But in this case only 1 g of fuel was
   burned anyway. So you're done.*

   (c)

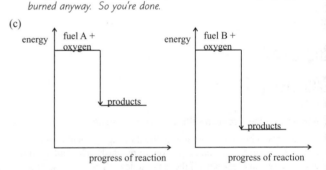

   *(1 mark for showing products lower than reactants, 1 mark if
   products are shown at a lower level for fuel B than for A.)*

Page 183

Revision Summary for Section 9

3) b)

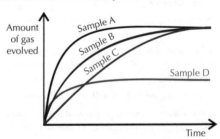

11) a) Mass of water heated = 116 g – 64 g = 52 g

   Temperature rise of water = 47 °C – 17 °C = 30 °C

   Mass of pentane burnt = 97.72 g – 97.37 g = 0.35 g

   So 0.35 g of pentane provides enough energy to heat up 52 g of water
   by 30 °C.

   It takes 4.2 joules of energy to heat up 1 g of water by 1 °C.

   Q = mcΔT.

   Therefore, the energy produced in this experiment is
   52 × 4.2 × 30 = 6552 J.

   So, 0.35 g of pentane produces 6552 J of energy... meaning 1 g of
   pentane produces 6552/0.35 = 18 720 J or 18.720 kJ

ANSWERS

Pages 188-189

Warm-Up Questions

) Which substances are present in the sample.

) A solution where the solvent is anything other than water.

) K^+

) blue

) Bubble the gas through limewater — the limewater will turn cloudy if it's carbon dioxide.

Exam Questions

Flame colour	Metal ion
green	Ba^{2+}
crimson	Li^+
yellow	Na^+
red	Ca^{2+}

(1 mark each)

(a) Calcium chloride *(1 mark for calcium, 1 mark for chloride)*.

(b) Potassium iodide *(1 mark for potassium, 1 mark for iodide)*.

(a) Add sodium hydroxide solution *(1 mark)*. A green precipitate indicates iron(II) ions *(1 mark)*.

(b) Add dilute hydrochloric acid and then barium chloride solution *(1 mark)*. A white precipitate indicates sulfate ions *(1 mark)*.

Add nitric acid and then silver nitrate solution *(1 mark)*.
A white precipitate indicates chloride ions *(1 mark)*.
A cream precipitate indicates bromide ions *(1 mark)*.
A yellow precipitate indicates iodide ions *(1 mark)*.

Al^{3+} *(1 mark)*
Cu^{2+} *(1 mark)*
Ca^{2+} *(1 mark)*

(a) carbon dioxide *(1 mark)*

(b) (i) Add sodium hydroxide solution *(1 mark)*. A brown precipitate indicates iron(III) ions *(1 mark)*.

(ii) $Fe^{3+}_{(aq)} + 3OH^-_{(aq)} \rightarrow Fe(OH)_{3\,(s)}$
(1 mark for all reactants and products correct, 1 mark for equation correctly balanced.)

Pages 197-198

Warm-Up Questions

) An unreactive gas, e.g. nitrogen.

) One mole of O_2 weighs $16 \times 2 = 32$ g.

) E.g. put an accurately measured volume of alkali in a flask. Add a few drops of indicator. Fill a burette with acid and record the volume. Add the acid to the alkali a bit at a time whilst giving the flask a regular swirl. Stop the reaction as soon as the indicator changes colour and record the volume of acid used to neutralise the alkali.

) $n = c \times V = 0.1 \times (25/1000) = 0.0025$ mol

) $M_r = (2 \times 23) + 12 + (3 \times 16) = 106$
Concentration in grams per $dm^3 = 0.025 \times 106 = 2.65$ g/dm^3

Exam Questions

(a) 4 *(1 mark)*

(b) 1 *(1 mark)*

(c) 3 *(1 mark)*

(d) 2 *(1 mark)*

2 (a) When heated, the electrons in an atom are excited *(1 mark)*, and release energy as light *(1 mark)*. The wavelengths emitted can be recorded as a line spectrum *(1 mark)*.

(b) Caesium and rubidium have different arrangements of electrons *(1 mark)*, so emit different wavelengths of light when heated *(1 mark)*. This means they have different patterns of wavelengths, and so different line spectrums *(1 mark)*.

(c) spectroscopy *(1 mark)*

3 (a) $n = c \times V$ *(1 mark)*
$n = 1.00 \times (30.3/1000) = 0.0303$ mol *(1 mark)*

(b) H_2SO_4 and NaOH react in a 1:2 ratio, so number of moles of H_2SO_4 $= 0.0303 \div 2 = 0.01515 = 0.0152$ mol (3 s.f.) *(1 mark)*

(c) $c = n \div V$ *(1 mark)*
$V = 25/1000 = 0.025$
$c = 0.01515 \div 0.025 = 0.606$ mol/dm^3 *(1 mark)*

4 (a) Any one from: e.g. it was an anomalous result/an outlier / the first titration is often a 'rough' titration and its result is not accurate *(1 mark)*.

(b) Moles of NaOH $= c \times V$
$= 0.1 \times (9.0/1000) = 0.0009$ mol *(1 mark)*
HA and NaOH react in a 1:1 ratio, so moles of HA $= 0.0009$ moles *(1 mark)*
Concentration of HA $= n \div V$
$V = 25/1000 = 0.025$
Concentration of HA $= 0.0009 \div 0.025 = 0.036$ mol/dm^3 *(1 mark)*

5 (a) Impurities in drugs can cause harm to the patients taking them *(1 mark)*.

(b) mass of drug $= c \times V$
$= 7.38 \times (25/1000) = 0.1845$ g *(1 mark)*
% purity $= 0.1845/0.23 \times 100 = 80.2\%$ *(1 mark)*

6

Test result	Gas
Turns litmus paper white	**Chlorine**
Makes a 'squeaky pop' with a lighted splint	Hydrogen
Relights a glowing splint	**Oxygen**

(1 mark for each correct answer).

Pages 204-205

Warm-Up Questions

1) It must be molten or dissolved in water.

2) Oxidation is loss of electrons and reduction is gain of electrons.

3) At the negative electrode.

4) Bromine.

5) The negative electrode.

Exam Questions

1 (a) (i) hydrogen *(1 mark)*

(ii) $2H^+ + 2e^- \rightarrow H_2$ *(1 mark)*

(b) (i) chlorine *(1 mark)*

(ii) $2Cl^- \rightarrow Cl_2 + 2e^- / 2Cl^- - 2e^- \rightarrow Cl_2$ *(1 mark)*

(iii) E.g. production of bleach / production of plastics *(1 mark)*.

(c) Sodium is more reactive than hydrogen, so sodium ions stay in solution *(1 mark)*. Hydroxide ions from water are also left behind *(1 mark)*. They combine to form sodium hydroxide in the solution *(1 mark)*.

2 (a) To lower the temperature that electrolysis can take place at *(1 mark)*. This makes it cheaper and easier *(1 mark)*.

(b) $Al^{3+} + 3e^- \rightarrow Al$ *(1 mark)*

(c) Oxygen is made at the positive electrode *(1 mark)*. The oxygen will react with the carbon in the electrode to make carbon dioxide *(1 mark)*. This will gradually wear the electrode away *(1 mark)*.

3 (a) (i) the pure silver strip *(1 mark)*

(ii) A solution containing silver ions, e.g. silver nitrate *(1 mark)*. The ions needed to plate the metal object come from the electrolyte *(1 mark)*.

(iii) E.g. silver is very expensive, so it's much cheaper to plate a brass cup with silver than it is to make the cup out of solid silver *(1 mark)*.

(b) It's a good conductor of electricity *(1 mark)*.

4 (a) When lead bromide is molten, its ions are free to move *(1 mark)*. It's these ions that conduct the electricity which allows electrolysis to work *(1 mark)*.

(b) At the cathode lead ions accept two electrons to become lead atoms / lead ions are reduced *(1 mark)*. At the anode bromide ions lose electrons and form bromine molecules *(1 mark)*.

(c) Cathode: $Pb^{2+} + 2e^- \rightarrow Pb$ *(1 mark)*
Anode: $2Br^- \rightarrow Br_2 + 2e^-$ *(1 mark)*

Page 206

Revision Summary for Section 10

5 The gas that the reaction gives off is hydrogen.

9 $R_f = 4.5 \div 12 = 0.375$

14 a) No. of moles NaOH = $0.2 \times (25 \div 1000) = 0.005$
$HCl + NaOH \rightarrow NaCl + H2O$,
so no. of moles HCl = 0.005
Concentration HCl (moles per dm^3)
= $0.005 \div (49 \div 1000)$
= **0.102 mol/dm³**

b) M_r HCl = $1 + 35.5 = 36.5$
mass = number of moles $\times M_r$
= $0.102 \times 36.5 = $ **3.72 g/dm³**

16 $(0.479 \div 0.5) \times 100 = 95.8\%$

Page 213

Warm-Up Questions

1) Water that doesn't form a lather with soap / Water that forms scale on the inside of pipes / Water that contains a lot of calcium and magnesium ions.

2) A furring on the inside of pipes, boilers and kettles that is mostly calcium carbonate.

3) temporary hardness and permanent hardness

4) By distillation (boiling water and condensing the steam).

5) enzymes

Exam Questions

1 (a) Any two of: e.g. Forms scum with soap. / Requires more soap for cleaning. / Forms scale on heating systems/kettles, etc. / May block pipes. *(1 mark each)*.

(b) Calcium ions are good for healthy teeth and bones. Studies have found that people who live in hard water areas are at less risk of developing heart disease than people who live in soft water areas. *(1 mark each)*

(c) When heated, the calcium hydrogencarbonate decomposes to form calcium carbonate *(1 mark)* which is insoluble *(1 mark)*.

(d) Adding sodium carbonate/washing soda *(1 mark)*. Using an ion exchange column *(1 mark)*.

2 (a) Filtration *(1 mark)* to remove solids *(1 mark)*.

(b) How to grade your answer:

0 marks: No advantages and disadvantages are given.

1-2 marks: Brief description of one advantage and one disadvantage is given.

3-4 marks: At least two advantages and two disadvantages are given. The answer has a logical structure and spelling, grammar and punctuation are mostly correct.

5-6 marks: The answer gives at least three advantages and three disadvantages. The answer has a logical structure and uses correct spelling, grammar and punctuation.

Here are some points your answer may include:

Advantages:

Water treated in this way has low levels of dissolved salts and no microbes.

Adding chlorine to water reduces the risk of getting a disease from drinking it.

This makes it suitable for humans to drink.

Adding fluoride to water reduces tooth decay.

Disadvantages:

Chlorine can react with other natural substances in the water to produce toxic by-products which some people think could cause cancer.

In high doses, fluoride can cause cancer and bone problems.

There are concerns about whether it's right to 'mass medicate' people as they have no say about what chemicals are used to treat their water.

3 (a) Detergent molecules have a hydrophobic tail and a hydrophilic head *(1 mark)*. The hydrophobic tail is attracted to the oil and surrounds it *(1 mark)* and the hydrophilic head is attracted to the water, pulling the oil away from the fabric *(1 mark)*.

Examiners would also have to accept an answer based on surfactant properties of detergents and the lowering of surface tension here — which is all getting a bit too fancy-pants in my opinion, but it just goes to show that there's often more than one correct way to explain things.

(b) Dry cleaning uses a solvent other than water / an organic solvent *(1 mark)*. The molecules in the dry cleaning solvent are strongly attracted to the molecules in the stain, pulling the stain apart *(1 mark)*.

Page 218

Warm-Up Questions

1) a) Ammonium phosphate is made by neutralising phosphoric acid with ammonia.

b) It can be used as a fertiliser.

2) eutrophication

3) exothermic

4) potassium hydroxide

Exam Questions

1 (a) An electrical cell that uses the energy from the reaction between a fuel and oxygen to produce electrical energy *(1 mark)*.

(b) A reaction which involves both reduction and oxidation *(1 mark)*.

(c) (i) Reduction, because the reactant is gaining electrons *(1 mark)*.

    (ii) $2H_2 + O_2 \rightarrow 2H_2O$
(1 mark for the equation, 1 mark if correctly balanced.)

(d) E.g. a hydrogen-oxygen fuel cell does not produce any pollutants *(1 mark)*.

(a) Fertilisers replace missing elements from the soil or add more of them *(1 mark)*. If plants don't get enough of these elements their growth and life processes are affected *(1 mark)*. So, using fertilisers increases crop yield *(1 mark)*.

(b) Any two of, e.g. nitrogen / phosphorus / potassium *(1 mark each)*.

(c) It contains nitrogen from two sources — the ammonia and the nitric acid *(1 mark)*.

(d) ammonium sulfate *(1 mark)*

(e) Algae living in the river water use the nitrates and phosphates to multiply rapidly, producing an algal bloom *(1 mark)*. This blocks off the light to the plants below *(1 mark)*. The plants cannot photosynthesise, so they have no food and they die *(1 mark)*. Aerobic bacteria feed on the dead plants and start to multiply *(1 mark)*. As the bacteria multiply, they use up all the oxygen in the water *(1 mark)*. As a result, other organisms in the river die *(1 mark)*.

Page 228

Warm-Up Questions

) E.g. ammonia / sulfuric acid / sodium hydroxide / phosphoric acid

) E.g. drugs / food additives / fragrances

) A life cycle assessment (LCA) looks at each stage of the life of a product to work out the potential environmental impact of each stage.

) Raw materials are the naturally occurring substances which are needed to produce a chemical.

Exam Questions

(a) (i) The atom economy of a reaction tells you how much of the mass of the reactants ends up as useful products *(1 mark)*.

    (ii) E.g. a process with a low atom economy produces a lot of waste, which has to be disposed of *(1 mark)*. The process will use up resources very quickly *(1 mark)*.

(b) A bulk chemical is one that is manufactured on a large scale *(1 mark)*.

(c) Feedstocks are the actual reactants needed for the process *(1 mark)*. Renewable feedstocks will not run out in the future *(1 mark)*.

(d) Any two of: e.g. governments monitor the use of chemicals *(1 mark)* — for example, how much of a chemical is used and any signs needed to warn the public of its use *(1 mark)*. / The way different chemicals are stored *(1 mark)* — for example, dangerous and poisonous chemicals should be locked away and stored in sealed containers or in well-ventilated store cupboards *(1 mark)*. / How chemicals are transported *(1 mark)* — for example, when transporting dangerous chemicals, lorries should display appropriate signs and emergency numbers that would be used if an accident happened *(1 mark)*.

(e) It is an assessment of anything in the process that could cause injury *(1 mark)*. It involves identifying hazards *(1 mark)*, assessing who might be harmed *(1 mark)* and deciding what action can be taken to reduce the risk *(1 mark)*.

(a) A sustainable process is one that meets people's needs today without affecting the ability of future generations to meet their own needs *(1 mark)*.

(b) How to grade your answer:

  0 marks: No discussion of sustainability.

  1-2 marks: Brief discussion of how waste products and product disposal affect sustainability.

  3-4 marks: Some discussion of at least three points regarding the sustainability of a product. The answer has a logical structure and spelling, grammar and punctuation are mostly correct.

  5-6 marks: Full discussion of at least five points regarding the sustainability of a product. The answer has a logical structure and uses correct spelling, grammar and punctuation.

Here are some points your answer may include:

Waste produces have to be carefully disposed of so they don't harm people or the environment.

Products are often disposed of in a landfill site at the end of their life. This takes up space and pollutes land and water. This is not sustainable.

Products might be incinerated, which causes air pollution. This is not sustainable.

Where possible, by-products can be used in other reactions. This increases the sustainability of the product.

If a reaction is exothermic, there may be waste heat. Heat exchangers can use excess heat to produce steam or hot water for other reactions — saving energy and money. This increases the sustainability of the product.

Pages 236-237

Warm-Up Questions

1) iron

2) They are recycled and used to produce more product.

3) vanadium pentoxide (V_2O_5)

4) distillation

Exam Questions

1 (a) (i) sugar *(1 mark)* $\xrightarrow{\text{yeast}}$ ethanol + carbon dioxide *(1 mark)*.

    (ii) zymase *(1 mark)*

    (iii) At lower temperatures the reaction slows down and at higher temperatures the enzyme is denatured *(1 mark)*. If the reaction solution is too acidic or too alkaline the enzyme will stop working *(1 mark)*.

    (iv) Oxygen converts the ethanol into ethanoic acid *(1 mark)* which lowers the pH and can stop the enzyme from working *(1 mark)*.

(b) (i) E.g. ethanol is put into a fractionating column *(1 mark)*. It is heated until boiling which allows the vapour to travel up the column *(1 mark)*. Only pure ethanol reaches the top of the column and condenses at the top, where it cools to create a liquid *(1 mark)*.

    (ii) When the concentration of ethanol reaches 10-20% the fermentation stops because yeast the is killed *(1 mark)*. So the fermented mixture is distilled to produce more concentrated ethanol *(1 mark)*.

(c) E.g. alcoholic drinks *(1 mark)*

2 (a) Ethane from crude oil or natural gas *(1 mark)* is 'cracked' (split) to form ethene and hydrogen gas *(1 mark)*.

(b) 300 °C *(1 mark)*, 70 atmospheres *(1 mark)*, phosphoric acid catalyst *(1 mark)*

(c) How to grade your answer:

0 marks: No discussion of sustainability.

1-2 marks: Brief discussion of the sustainability of the process.

3-4 marks: Some evaluation of at least three points regarding the sustainability of the process. The answer has a logical structure and spelling, grammar and punctuation are mostly correct.

5-6 marks: Full evaluation of at least five points regarding the sustainability of the process. The answer has a logical structure and uses correct spelling, grammar and punctuation.

Here are some points your answer may include:

Crude oil and natural gas are non-renewable and will run out. This is not sustainable.

Cracking ethane has a fairly high atom economy as the only waste product is hydrogen. Reacting ethene has an even higher atom economy as ethanol is the only product. This increases the sustainability of the product.

The only waste is the hydrogen gas produced by cracking ethane. It can be reused to make ammonia in the Haber process. This increases the sustainability of the product.

Energy is needed to maintain the high temperature and pressure used.

The reactions involved do not produce any waste products that directly harm the environment. But, crude oil can harm the environment, e.g. through oil spills.

Manufacturing ethanol from ethene and steam is continuous and quick and the raw materials are fairly cheap — but it won't stay that way once crude oil starts to run out

3 (a) The Contact process *(1 mark)*.

(b) (i) A higher temperature increases the rate of reaction *(1 mark)*.

(ii) 1 atmosphere/atmospheric pressure *(1 mark)*.
There are fewer moles of product than there are of reactant so a high pressure would give more product *(1 mark)*, but high pressures are expensive to maintain/ are not really needed, as the equilibrium is already on the right *(1 mark)*.

4 (a) (i) E.g. crude oil/natural gas *(1 mark)*

(ii) air *(1 mark)*

(b) $3H_2(g) + N_2(g) \rightleftharpoons 2NH_3(g)$
(1 mark for correct formula, 1 mark if correctly balanced, 1 mark for correct state symbols).

(c) A high temperature reduces the equilibrium yield but increases the rate of the reaction *(1 mark)*. If the temperature was any lower, the product would be formed too slowly *(1 mark)*.

Remember, it's better to get a yield of 10% after 20 seconds than a yield of 20% after 60 seconds.

Page 244

Warm-Up Questions

1) refluxing, distillation, purification, drying

2) The higher the melting point the stronger the forces holding the polymer chains together.

3) crosslinks

4) E.g. the starting materials and the reaction conditions.

5) A shape memory alloy (about half nickel and half titanium) / a smart material

Exam Questions

1 (a) (i) nm *(1 mark)*

(ii) surface area *(1 mark)*, volume *(1 mark)*

(b) (i) The CNTs provide strength *(1 mark)* and lightness/low density *(1 mark)*.

(ii) fullerenes *(1 mark)*

(c) Any one of: e.g. in computer chips / in sensors / as catalysts / delivering drugs / in cosmetics / in lubricants *(1 mark)*.

2 (a) B *(1 mark)*

(b) C (accept A) *(1 mark)*

(c) A *(1 mark)*

Page 248

Warm-Up Questions

1) Any two of, e.g. they're non-toxic / non-flammable / chemically inert / insoluble in water / they have low boiling points.

2) Any one of, e.g. aerosol propellants / fridge coolants / air-conditioning systems.

3) An atom or molecule with an unpaired electron.

4) $CCl_2F_2 \xrightarrow{\text{UV}} CClF_2\cdot + Cl\cdot$

5) They are unreactive and can exist in the atmosphere for long periods before starting to react.

Exam Questions

1 (a) They are unreactive/inert, but the strong ultraviolet light in the stratosphere has enough energy to break up the molecules *(1 mark)*.

(b) (i) $O_3 + Cl\cdot \rightarrow O_2$ *(1 mark)* $+ ClO\cdot$ *(1 mark)*

(ii) $ClO\cdot$ *(1 mark)* $+ O_3 \rightarrow 2O_2 + Cl\cdot$ *(1 mark)*

(c) It is a chain reaction *(1 mark)*. The Cl• free radical is a product/ is recycled and can react with more ozone *(1 mark)*.

(d) alkanes *(1 mark)* and HFCs/hydrofluorocarbons *(1 mark)*

2 (a) Ozone is a form of oxygen with three oxygen atoms per molecule *(1 mark)*. The ozone layer is a layer of ozone that is in the stratosphere/part of the upper atmosphere *(1 mark)*.

(b) E.g. Antarctica *(1 mark)*

(c) The answer should contain three of the following points:
ozone protects against harmful UV radiation
ozone levels have fallen
incidence of skin cancer has increased
but other factors, e.g. more holidays, may play a part
(1 mark per point; maximum 3 marks)

If the question says 'discuss', try to put across more than one point of view.

3 B *(1 mark)*.

Index

Index

Index

CGP

Make sure you're not missing out on another superb CGP revision book that might just save your life...

...order your **free** catalogue today.

CGP customer service is second to none

We work very hard to despatch all orders the **same day** we receive them, and our success rate is currently 99.9%. We send all orders by **overnight courier** or **First Class** post.
If you ring us today you should get your catalogue or book tomorrow. Irresistible, surely?

- Phone: 0870 750 1252   (Mon-Fri, 8.30am to 5.30pm)
- Fax: 0870 750 1292
- e-mail: orders@cgpbooks.co.uk
- Post: CGP Orders, Broughton-in-Furness, Cumbria, LA20 6BN
- Website: www.cgpbooks.co.uk

...or you can ask at any good bookshop.